Crop rotation studies on the Canadian prairies

C.A. Campbell and R.P. Zentner
Research Station
Swift Current, Sask.

H.H. Janzen
Research Station
Lethbridge, Alta.

K.E. Bowren (retired)
Research Station
Melfort, Sask.

Research Branch
Agriculture Canada

Publication 1841/E
1990

Available in Canada through
Authorized Bookstore Agents
and other bookstores
or by mail from

Canadian Government Publishing Centre
Supply and Services Canada
Ottawa, Ontario K1A 0S9

Catalog No. A53-1841-1990E
ISBN 0-660-13422-5

Canadian Cataloguing in Publication Data

Crop rotation studies on the Canadian prairies

(Publication ; 1841)

Includes bibliographical references.
Cat. no. A53-1841/1990E
ISBN 0-660-13422-5

1. Crop rotation–Prairie Provinces. I. Campbell, C. A. (Constantine A.), 1934– . II. Canada. Agriculture Canada. III. Series: Publication (Canada. Agriculture Canada). English ; 1841.

S603.C7 1990 631.5'82 C90-099101-1

Staff editor: Jane T. Buckley

To the scientists, who had the foresight to plan and implement these studies, the technicians, who diligently maintained them, analyzed them, and kept the records, and the managers, who supported their continuance through periods of reduced resources and sometimes mixed signals on research priorities.

Recommendations for pesticide use in this publication are intended as guidelines only. Any application of a pesticide must be in accordance with directions printed on the product label of that pesticide as prescribed under the *Pest Control Products Act*. **Always read the label**. A pesticide should also be recommended by provincial authorities. Because recommendations for use may vary from province to province, your provincial agricultural representative should be consulted for specific advice.

Cover illustration

Crop rotation studies at Indian Head, Sask., have been carried on for the past 100 years.

CONTENTS

ABBREVIATIONS

The following abbreviations are used in this text:

A	alfalfa
B	barley
Br	bromegrass
Can	canola
CL	clay loam
Co	corn
DM	dry matter
ET_p	potential evaporation
F	fallow
Flx	flax
FS	fallow substitute crops
H	hay
GM	sweetclover green manure
G_N	grain nitrogen
G_P	grain phosphorus
GDD	growing degree days
M_e	moisture use efficiency
M_{gs}	growing season precipitation
M_{hv}	harvest soil moisture
M_s	soil moisture
M_{sp}	spring soil moisture

M_u	moisture use
O	oats
Pe	peas
Po	potatoes
Ry	rye
Sor	grain sorghum
SC	sweetclover hay
SL	sandy loam
S_N	straw nitrogen
S_P	straw phosphorus
W	spring wheat
WW	winter wheat
Y_g	yield of grain
Y_s	yield of straw
Y_{cv}	variability of yield

PREFACE

Long-term crop rotation studies require tremendous commitment of resources (money, land, personnel, and equipment) and are often difficult to terminate. Administrators, especially in times of restraint, are usually under pressure to reassess the need for maintaining these costly studies. Such decisions are best made after assessment of the status of current studies and the future benefits that they may provide. Thus, a committee was set up by the director general of the Research Branch for Agriculture Canada Western Region, to summarize the findings of Agriculture Canada's recent rotation studies in the prairies, to write a bulletin, and to make recommendations so that intelligent decisions regarding the fate of these studies can be made. The committee consisted of the four authors plus the following: Mr. S.A. Brandt (Scott) and Drs. C.A. Grant (Brandon), G.P. LaFond (Indian Head), D.W. McAndrew (Vegreville), Y.K. Soon (Beaverlodge), and L. Townley-Smith (Melfort).

This publication is the product of this exercise. It is targeted primarily at the scientific community, but extension personnel will also find it to be a useful reference. It deals mainly with data gathered since the early 1960s.

Only dryland studies are discussed in this book; readers interested in Agriculture Canada's irrigated rotations should refer to the technical bulletin *Ten-year irrigated rotation U. 1911–1980* (Dubetz 1983).

A concise version of this book (Agric. Can. Publ. 1839/E), entitled *Benefits of crop rotation for sustainable agriculture in dryland farming*, presents a useful summary of the main conclusions arrived at herein. Publication 1839 is targeted primarily at producers and extension workers and is available in English and French from Communications Branch, Agriculture Canada, Ottawa, Ont. K1A 0C7.

ACKNOWLEDGMENTS

Drs. H.W. Cutforth, R.D. Tinline, J.R. Byers, and L.D. Bailey, of Agriculture Canada research stations at Swift Current, Saskatoon, Lethbridge, and Brandon, respectively, wrote the review sections of the book dealing with climate and soils, plant diseases, insects, and legumes, pulse, and forages, respectively. Their excellent contributions are greatly appreciated.

All members of the committee cited in the Preface supplied information from their respective research stations and reviewed early versions of the manuscript. Their assistance and encouragement were invaluable and are appreciated.

The following people reviewed the manuscript and provided valuable suggestions: Drs. J.W.B. Stewart (University of Saskatchewan, Saskatoon); L.D. Bailey (Agriculture Canada, Brandon); C.V. Cole (U.S. Department of Agriculture, National Resource Ecology Laboratories, Fort Collins, Colorado State University); Paul L. Brown (Montana State University); and Messrs. W. Earl Johnson (Retired, Saskatchewan Department of Agriculture, Regina) and Roy Button (Saskatchewan Rural Development).

INTRODUCTION

A crop rotation is a planned sequence of crops grown in recurring succession on the same area of land. In western Canada, crop rotation effects on crop yields and productivity have received more attention from researchers than any other topic. This interest exists not only because crop rotation studies play an important role in ensuring optimum productivity, but also because complex agronomic interactions are intrinsic to these lengthy studies.

The benefits of crop rotations in agricultural production include improved resistance to soil erosion and degradation, improved soil fertility, soil tilth, enhanced aggregate stability, increased availability of stored moisture, improved pest control, reduction of allelopathic or phytotoxic effects, enhanced economic stability, and more even distribution of work load.

The producer's choice of crop rotations is influenced by three sets of factors. The first set consists of physical factors and includes soil characteristics, climatic constraints, intercrop antagonisms and synergisms, effects of soil and environmental quality, and the incidence of pests. These factors determine crops that can be grown, substitution possibilities, and expected yields.

The second set of factors consists of economic considerations. These include amount and seasonality of resources, expected prices for products, costs of inputs and credit, marketing opportunities, agricultural policies and programs, tax considerations, financial position of the farm, availability of equipment and labor, and ability of the farm to withstand major fluctuations in income. These factors provide the criteria on which to base rational decisions by weighing the relative advantage of each crop, agronomic consideration, and resource service in relation to financial goals of the farm. The cropping program that is optimal for an individual farm depends on the relative importance attached to these criteria and the management expertise available.

The third set of factors includes the decision-making and organizational abilities of farm managers, which are based on knowledge, skills, management ability, and attitudes towards risk. These factors determine or govern the degree of success of farm managers in processing information and in choosing and directing the optimal cropping program for their particular farm.

In studying crop rotations, researchers are confronted by numerous problems and challenges. They must constantly be aware of an array of soil and crop variables and a multiplicity of interactions among them. Furthermore, some effects do not become apparent for years, whereas others become clear early on. Consequently, field experiments must be monitored regularly throughout the year and over many years if reliable and transferable information is to be obtained. This duration may necessitate the use of analyses that account for temporal trends in weather, management, soil productivity, cultivars, and scientists. Conclusions may be difficult to draw and many require input from experienced researchers. These studies spawn considerable data, which, even in this computer age, require dedication and commitment on the part of those whose job includes their summarization and critical analysis.

Crop rotation studies have been conducted in western Canada since the 1890s. Most studies were carried out by Agriculture Canada, but a limited number have been conducted by universities (Poyser et al. 1957, Robertson 1979, Robertson and McGill 1983). Hopkins and Barnes (1928) and Hopkins and Leahey (1944) first attempted to summarize the results of experiments conducted by Agriculture Canada. Later, Brown (1964) of the U.S. Department of Agriculture included a chapter on the western Canadian experience to 1960 in his publication on legumes and grasses. Ripley (1969) presented a comprehensive summary of crop rotations carried out by Agriculture Canada up to 1965, including a thorough review of the pertinent literature.

REVIEW OF CROPPING CONDITIONS IN WESTERN CANADA

Several major agronomic factors may either influence crop rotations (e.g., climatic and edaphic factors) or be influenced by rotations (e.g., weeds, diseases, insects, soil quality, and net returns). This section provides background information on some of these factors to facilitate the understanding of subsequent discussions.

Arable soils of the Prairie Provinces range in type from Brown and Dark Brown in southern Saskatchewan and Alberta, to Black in Manitoba and Dark Gray soils in the central prairies and Gray (Luvisolic) soils in the North (Fig. 1). In the United States system of soil taxonomy, these soils are called Aridic Haploboroll, Typic Boroll, Udic Boroll, and Alfisols, respectively. The native vegetation of the dry Brown and Dark Brown soils consists mainly of xerophytic and mesophytic grasses and forbs. The Black soils occur in the fescue prairie–aspen grove (parkland) and true prairie grasslands, whereas the Dark Gray soils are located in the transitional areas of grassland and forests. The Gray soils developed under mixed deciduous and evergreen forests on high basic, mineral parent materials in subhumid to humid, mild to very cold climates.

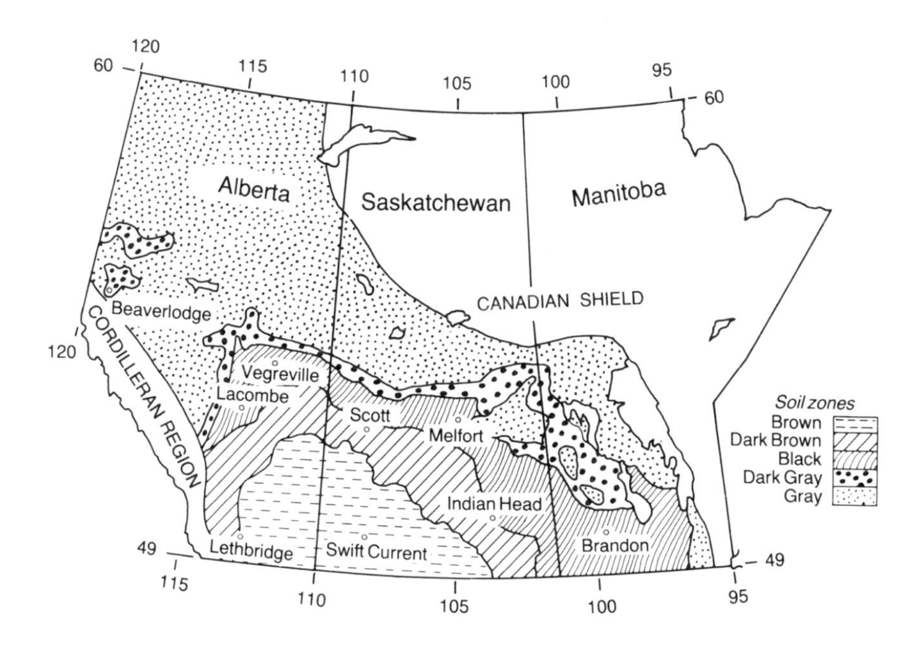

Fig. 1 Location of research stations and major soil zones of the prairie region.

CLIMATE

Mean monthly precipitation, temperature, and estimated potential evaporation (ET_p) for research stations that have conducted rotation studies are shown in Fig. 2. The stations are located within the major prairie soil zones. Within each soil zone there are differences or gradations in climate so that no one location can suitably represent all areas of a zone. For example, Swift Current, although in the Brown soil zone, represents conditions much closer to Dark Brown soils and does not truly represent drier conditions in areas around Burstall, Bracken, and Climax. Similarly, Scott in the Dark Brown soil zone is much wetter than Estevan in the same soil zone, and Black soils at Melfort differ greatly from those at Indian Head. This fact must be remembered both in data interpretation and extrapolation.

Throughout the arable prairies, the warmest and coldest months are July and January, respectively. June is the month with greatest precipitation and July the month with highest ET_p. Daily and seasonally, temperatures fluctuate widely. Differences between the warmest and coldest months vary between 29 and 39°C (Table 1), differences increasing from west to east. Daily maximum–minimum temperature differences of 20–25°C are not unusual. About 50% of the annual precipitation falls from May to September, with about 30% falling as snow during the winter months. Snow acts as an important source of soil moisture, insulates the soil, and protects it against erosion and drying. Potential evaporation generally exceeds precipitation thus resulting in a moisture deficit during the months April to September, whereas precipitation generally exceeds ET_p during the winter months, November to March.

In general, annual precipitation increases from <350 mm in the Brown soil zone to >475 mm in parts of the Black and Gray soil zones (Table 2). However, potential evaporation decreases from the Brown to the Gray soil zones. Hence, annual moisture deficits decrease from about 400 mm in the Brown soil zone to little or none in the Gray soil zone. Mean annual temperatures are generally higher in the Brown and Dark Brown than in the Black and Gray soil zones. Windspeeds in the chinook area of southern Alberta (Lethbridge) and southwestern Saskatchewan (Swift Current) are substantially greater than elsewhere on the prairies.

Mean annual temperatures, frost-free days (>0°C), and annual growing-degree days above 5°C (GDD) increase from north to south (Table 1). Eastward from the foothills of Alberta, winters are usually colder and summers warmer. Thus, from west to east, GDD and frost-free days generally increase slightly whereas the mean annual temperature decreases slightly.

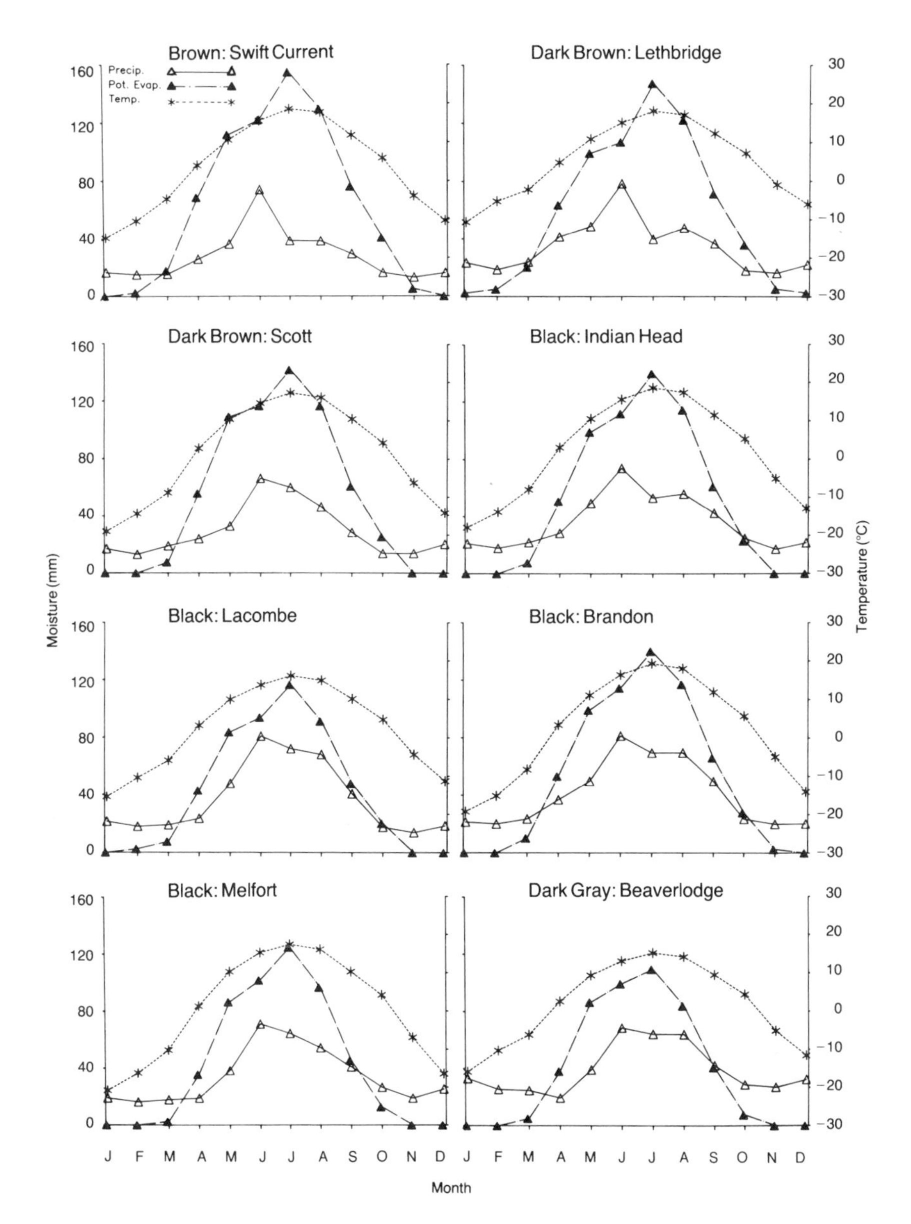

Fig. 2 Mean monthly climatic data for Agriculture Canada research stations at which studies were carried out, 1951–1980.

Table 1 Annual growing-degree days >5°C, average frost-free days (>0°C), and mean annual temperature (1951–1980) for selected Agriculture Canada research stations on the prairies

Location	Growing-degree days (>5°C)							Frost-free days (>0°C)	Mean annual temp. (°C)
	April	May	June	July	August	Sept	Annual		
Alberta									
Beaverlodge	25	145	242	317	285	149	1221	105	1.6 (31)†
Lacombe	35	160	262	345	305	162	1334	99	2.1 (32)
Lethbridge	60	184	300	404	372	221	1689	116	5.0 (29)
Saskatchewan									
Scott	37	174	286	377	343	170	1442	97	1.0 (36)
Melfort	40	175	308	385	345	169	1468	93	0.3 (38)
Swift Current	53	182	316	418	393	214	1675	117	3.3 (33)
Indian Head	44	185	319	421	386	201	1633	110	2.0 (37)
Manitoba									
Brandon	47	196	339	441	399	207	1705	104	1.9 (39)

† The figures in parentheses are the temperature difference between the warmest and coldest month.

Table 2 Mean annual windspeed, temperature, and precipitation (1951–1980), calculated potential evaporation (ET_p) (1931–1960), and moisture deficit at selected Agriculture Canada research stations

Location	Wind† ($km \cdot h^{-1}$)	Temp. (°C)	Precip. (mm)	ET_p (mm)	Deficit†† (mm)
Brown soil zone					
Swift Current	22.9	3.3	334	729	395
Dark Brown soil zone					
Lethbridge	20.4	5.0	413	681	268
Scott	14.5	1.0	355	635	280
Black soil zone					
Indian Head	15.8	2.0	427	607	180
Lacombe	10.9	2.1	443	508	65
Brandon	16.3	1.9	485	630	145
Melfort	15.4	0.3	411	506	95
Dark Gray soil zone					
Beaverlodge	12.2	1.6	467	470	3

† 10 m above ground.
†† Deficit = ET_p − Precipitation.

SOIL ZONES

The Brown soil zone occupies the south-central region of the plains area (Fig. 1). The average depth of the surface layer is 12.5 cm and the organic matter content of the surface 30 cm of soil is about 2%. About 80% of these glaciated soils are of a medium (loam) texture. Topography varies from nearly level to very hilly. Wind erosion is a serious problem. In this soil zone, moisture deficit is the major consideration in developing farming and tillage systems with long-term production and economic capabilities.

In the Dark Brown soil zone the depth of the surface horizon averages about 17.5 cm. Organic matter content of the surface 30 cm is about 4%. As in the Brown soil zone, medium-textured soils predominate but the land is more nearly level. Here too, moisture deficit dictates that water conservation must always play a major role in the design of crop rotation and tillage systems. However, the lower moisture deficit results in greater flexibility in choice of crop rotation than is possible on the Brown soils. Wind erosion is less of a problem compared with the Brown soil zone because of the denser vegetative cover. In both the Brown and Dark Brown soil zones, salinity is a widespread and increasing problem (Campbell et al. 1986).

In the Black soil zone the depth of the surface horizon averages 20–25 cm, and the soil contains about 7% organic matter. The soil is mostly medium textured; the land is mainly level to gently rolling. Wind erosion is not generally a problem except in localized areas where large fields are summer-fallowed.

The Gray Black and Gray Luvisolic forest soils occur north of the Black soil zone and include the Peace River region of Alberta. Generally, they have a thin layer of dark-colored humus, about 5 cm deep over a layer of gray-colored soil. The organic matter content of this grayish layer is generally low. With sufficient applications of nitrogen, phosphorus, and sulfur fertilizers these soils have good productivity. Some acid soils (pH < 6.0) occur within this region, mainly in the Peace River area. The Gray Black and Gray Luvisolic soils have organic matter contents ranging from as low as 1% to as high as 10% and also include organic soils with 30% or more organic matter. The loose, powdery nature of the Ap horizon of soils with low organic matter (2–3%) results in poor structure, which results in puddling and crusting following heavy rains. These conditions may result in poor or uneven plant emergence. This zone also presents a wider range of parent material and soil temperature variation than other regions. Because of the reduced soil moisture deficit, sandy soils may be improved from native grassland status to soils that are highly productive for grain production. Early fall frost is one of the most limiting factors in many parts of this zone; again sandy soils present fewer problems than heavy-textured soils in this respect. Wind erosion is not a serious problem but water erosion is a major concern. In the Black and Gray soil zones, salinity is an increasing problem but is not as widespread as in the drier soil zones.

Solonetzic soils constitute a significant proportion of soils in Alberta and, to a lesser, extent in Saskatchewan. For example, about 30% of all arable land in Alberta is solonetzic. These soils exist in complex association with the normal soils in all soil zones. Often referred to as "burnout," "blowout," or "gumbo" soils, they are characterized by a tough, impermeable hardpan located 5–30 cm below the soil surface. The hardpan restricts root growth and water penetration; consequently crop yields vary considerably from year to year depending on rainfall distribution (Hermans 1977).

CROP SELECTION AND MANAGEMENT

Most farms in the Prairie Provinces are family operations, owned or rented by the operators. They are large, highly mechanized, and highly seasonal in terms of inputs and products. Income is often dependent on one crop or enterprise, and production is risky and subject to many dynamic forces. In all soil zones, spring wheat is the principal crop, although mixed farming is more prevalent in the Black, Dark Gray, and Gray soil zones.

Cropping

On the Canadian prairies, producers have historically selected crop rotations that include high proportions of wheat and summer fallow. The predominant fallowing frequency ranges from once every 2 years in the Brown soil zone to once every 4 years or more in the Black soil zone, in direct relation with moisture availability. Although wheat is the major crop, barley and canola occupy significant proportions of the seeded area in the Dark Brown and Black soil zones (Zentner et al. 1986). In the Gray Luvisols and Dark Gray Luvisolic soils, mixed farming is common; rotations involve a high proportion of forage legumes and grasses for hay and pasture (Hoyt et al. 1978).

Of about 30 000 000 ha of cultivated lands in the three Prairie Provinces (Campbell et al. 1986), about 50% is located in the Black and Gray soil zones, 27% in the Dark Brown zone, and 23% in the Brown zone. Fallow land increased from 10 000 000 ha in the late 1950s to 11 500 000 ha in the early 1970s, but has since decreased to 9 000 000 ha, with most of the decrease occurring between 1980 and 1985. From 1957 to 1984, the area of cultivated land increased by 5 000 000 ha; 80% of this increase resulted from breaking of new land and the cultivation of pasture or other grassland areas. Fallow area has remained relatively constant in the Brown soil zone as has cropped land. The Dark Brown soils have gained 400 000 ha of cultivated land, but most of the increase in cultivated land on the prairies has occurred in the Black and Gray soil zones (3 000 000 ha). Much of the newly broken land occurred in the Gray Luvisolic soils of the northern fringes of the parkland.

Soil quality

Soil quality refers to the ability of a soil to sustain, accept, store, and recycle nutrients, water, and energy (Anderson and Gregorich 1984). Good soil quality is not always compatible with the realities of short-term economic circumstances. Not surprisingly then, soil quality has declined at an unacceptable rate on the prairies, which has resulted in a strong call for immediate action to reverse this trend (Sparrow 1984). Erosion, salinization, acidification, and declining levels and quality of organic matter have been identified as the major sources of soil degradation in western Canada (Coote et al. 1981). Although many of these processes occur naturally, our cultural and management practices have accelerated their rates to the point at which the ability of the soils to produce food on a sustained basis is uncertain. Further, the federal government's price and income policies and programs often create incentives for cropping systems that contribute to soil degradation (e.g., quotas), which influence soil quality or productivity.

Soil degradation reduces soil productivity by causing loss of plant nutrients, reduction in moisture-holding capacity, deterioration of soil structure, poor conditions in the seedbed, reduction in microbial activity within the soil, and interference with the osmotic and biosynthetic processes of plants (National Soil Erosion–Soil Productivity Research Planning Committee 1981). The aggregate economic benefits from arresting further soil degradation in the Prairie Provinces have been estimated to exceed $3.2 billion through the next 18 years alone (Prairie Farm Rehabilitation Administration 1982). Crop rotations and proper management can be used to alleviate these problems.

Fertilizer use

Prior to 1960, the high fertility levels of prairie soils at the time of original breaking and the frequent use of summer fallow to build up plant-available nutrients resulted in little need or use of commercial fertilizers for annual crop production (Kraft 1980). Since 1957, the use of nitrogen (N) and phosphorus (P) fertilizers has increased rapidly, except for the years 1969–1972 when unfavorable economic conditions existed (Campbell et al. 1986).

The increased use of fertilizers on the prairies has resulted partly from more stubble cropping and partly from removal of nutrients by crops and general soil degradation (Campbell et al. 1986).

Tillage

Mechanical tillage is an important component of most cropping systems and provides several benefits. First, it prepares the seedbed. A good seedbed for cereals, oilseeds, and other small-seeded crops provides the seed with sufficient water for germination, has a good mix of soil aggregates to prevent or minimize soil erosion, and has the capability to hold water for growth and development of plants. Second, tillage controls weeds not only prior to seeding, but also during a fallow phase of a crop rotation and in the inter-row areas of row crops. Third, tillage is used to incorporate herbicides and other pesticides into the soil.

The benefits of soil tillage are not achieved without potential loss of soil productivity. First, tillage can enhance moisture loss from soil thereby reducing germination and subsequent vigor of crops. Second and more important, excessive tillage can greatly increase the susceptibility of soil to erosion by destroying the soil's aggregation and by reducing the amount of crop residues retained on the soil's surface. Third, tillage can accelerate the loss of organic matter, a vital constituent of soil, by increasing aeration and temperature within the soil. These potentially degradative influences of tillage on soil productivity merit consideration in the development of crop management strategies. In many cases, the degree of soil degradation arising from tillage can be minimized by appropriate selection of tillage implements (Anderson 1961, 1967).

An important consideration in the development of effective yet nondegradative tillage is optimizing its timing. Several researchers in western Canada have examined the relative merits of fall and spring tillage. The results of these studies vary primarily as a function of soil moisture regimes. In the Brown and Dark Brown soil zones, where moisture limits crop production, fall tillage may be beneficial only if there is a heavy infestation of weeds. In a Dark Brown soil at Scott, Matthews (1949) observed higher yields of wheat after spring tillage than after various treatments using fall tillage. In Black soils, fall tillage has no beneficial effect on storage of winter precipitation (Janzen et al. 1960, Michalyna and Hedlin 1961, Nuttall et al. 1986). Indeed, fall tillage may reduce moisture conservation by disturbing standing stubble, which diminishes its effectiveness as a snow trap (Staple et al. 1960). Because of this effect and the potential increase in the soil's erodibility, fall tillage in the Dark Brown and Brown soil zones is discouraged. In contrast, fall tillage may offer some advantage in the more humid regions where moisture is rarely limiting and where crop residues may pose a problem to preparation of the seedbed (Lal and Steppuhn 1980). However, even in the Black and Gray soil zones, fall tillage may not always result in increased yields. Nuttall et al. (1986) obtained best results when straw was chopped and left on the soil surface without tillage, to hold snow. Therefore, fall tillage appears justified only where the accumulation of crop residues is excessive, regardless of soil zone (Bowren and Dryden 1971).

Tillage is the traditional means of controlling weeds during the fallow year of cereal cropping systems. Tillage should commence early in the growing season to minimize consumption of nutrients and moisture by weeds and to prevent maturation of weed seedlings. Tillage should be discontinued in late summer; subsequent weed growth will be killed by frost prior to maturation yet will serve as an effective soil cover. The degradative effects on soils of tillage during the fallow period can be minimized either by use of implements that result in little reduction of trash and disturbance of soil (Anderson 1961, 1967) or by judicious replacement of one or more tillage operations with applications of nonselective herbicide (Lindwall and Anderson 1981). Effective fallowing practices are measured not only in terms of their efficacy in weed control, but also in their capacity for soil conservation.

Weed control

Weeds are a serious problem to crop production in western Canada. When allowed to grow unchecked, they reduce crop yield through competition for soil water and nutrients and jeopardize quality and harvestibility of the crop. As a result, considerable effort is expended in the development of strategies for the eradication or control of weeds.

Weed control is particularly important during the summer-fallow phase of crop rotations. Research has demonstrated that the timing of tillage, the traditional method of weed control, strongly influences subsequent crop yields. Cultivation of stubble during the first fall of the fallow period has shown little benefit (Anderson 1967), but early tillage in spring (before May 15) minimizes use by weeds of water and nutrient reserves (Bowren 1977, Austenson 1978). In southwestern Saskatchewan, delaying initial tillage until 1 June resulted in a 10% reduction in yield; further delay of tillage until 1 July reduced subsequent yields by 20%. Similar effects were observed in other parts of Saskatchewan, though the benefits of early tillage were not as pronounced under conditions of above-normal precipitation during the growing season (Austenson 1978).

Growth of weeds during the fallow phase can be controlled effectively by applying herbicides (Anderson 1971). Research at Swift Current has shown that wheat yields after chemical fallow were comparable to those after tilled fallow. In chemical-fallow systems, potential problems are the control of volunteer cereals and retention of herbicides in the soil. Application of soil-applied herbicides in late fall, however, has been shown to maintain residual control of most volunteer crops and weeds during the fallow period yet permit dissipation of the herbicide prior to establishment of the crop in the following year. The efficacy of this method and its residual effects on subsequent crops vary depending on soil type.

Weed control in fallow, by tillage or herbicides, has some inherent limitations. As a result, the best strategy for weed control may involve a combination of the two methods. For example, application in late fall of herbicides such as 2,4-D may allow delay of initial tillage until mid June of the following year (Molberg et al. 1967, Anderson 1971, Bowren 1977) and may reduce the number of tillage passes required for effective control of weeds.

The primary means of controlling weeds in a crop is the application of selective herbicides. Many herbicides are available for weed control in all common crops. Effective and economical use of these products, however, depends on factors that include application at appropriate growth stage, proper selection and adjustment of sprayer equipment, care in application to avoid overlap and drift, and, for soil-applied herbicides, effective incorporation.

A possible limitation to continued widespread use of chemical herbicides is their perceived harmful effects on the environment over the long term. Some studies have demonstrated complete, rapid dissipation of many herbicides following application to the soil. For other herbicides, carryover is observed. Prolonged persistence of herbicides or herbicide degradation products in the soil is of concern in western Canada because microbial decomposition rates are greatly slowed by cool soil temperatures during most of the year. The effects on soil processes of repeated applications of herbicide have not been fully ascertained, though Biederbeck et al. (1987) demonstrated no deleterious effects on soil biota after 40 years of using 2,4-D.

The degree and composition of weed infestations can be controlled to some extent by adjustment of crops within a rotation. For example, barley competes relatively strongly whereas flax competes poorly with weeds. Grasses and legumes grown for forage generally reduce infestations in crop rotations but may pose problems by regrowth in subsequent grain crops. Fall-seeded crops, because of their early regrowth in spring, compete well with weeds when environmental conditions are favorable. However, after severe winterkill, they may suffer severe infestations of weeds. Furthermore, their early maturity may allow perennial weeds to become well established late in the growing season, which then causes problems for subsequent crops. Continuous cropping of annual cereals may lead to serious infestation of grassy weeds such as wild oats, quack grass, green foxtail, foxtail barley, barnyard grass, Persian darnel, and, in the case of winter wheat, downy brome. Effective weed control, therefore, can be facilitated by selection of crop sequences that use competitive crops to control particular weeds, allow for application of appropriate selective herbicides to control specific weeds, and maintain a diversity of crops.

The type of weed control adopted for specific production systems should provide high efficacy of control, maintain soil and environmental quality, and be economically feasible. A particular strategy can usually best be achieved through the complementary use of tillage, herbicide, and cropping sequence.

Plant diseases

Rotation of susceptible and resistant crops is one of the oldest practices used to control disease. It remains an important practice against many diseases, particularly those for which more specific controls, such as host resistance or chemical methods, are unavailable.

The success of rotation in disease reduction is contingent upon many factors, which include the ability of a pathogen to survive in the absence of its host and the host range of a pathogen. Pathogens that live indefinitely in the soil are less likely to be curtailed by rotation of crops than those that can survive for only brief periods apart from their hosts. Similarly, pathogens that have a wide range of hosts are less amenable to control by crop rotation than those with a narrow range. Transmission of pathogens via seeds, the presence of susceptible volunteer crops and weeds that harbor the pathogens, and the distribution of pathogens by wind and other agents may negate benefits derived from crop rotation. For example, rotation is ineffective for control of rusts in small grain cereals because the rust fungi do not overwinter in western Canada. Inoculum from the south is disseminated into the area by wind in summer, which circumvents protection by crop rotations.

Typically, crop rotation is used in conjunction with other cultural practices such as tillage, fertilizer, and weed control. Reduction in disease may relate to various factors including nutritional and biological ones in addition to the decline of inoculum of a pathogen in the absence of the susceptible host.

Crop rotation is a recommended control practice for a number of leaf blights of small grain cereals. Shaner (1981) in reviewing the effect of environment on some major leaf diseases of cereals indicated that the responsible pathogens are not carried long distances by wind. The pathogens overwinter in crop debris and on seed. Net blotch (*Pyrenophora teres* Drechs.), scald (*Rhynchosporium secalis* (Oud.) J.J. Davis), and speckled leaf blotch (*Septoria passerinii* Sacc.) of barley, and tan spot (*Pyrenophora tritici-repentis* (Died.) Drechs.) and septoria blotches (*Mycosphaerella graminicola* (Fuckel) Schroeter and *Phaeosphaeria nodorum* (Müller) Hedjaronde) of wheat were some of the diseases he discussed. A break of 2–3 years between susceptible crops is usually considered sufficient to nullify inoculum in the crop debris. Sanitation, to eliminate volunteer host plants that may harbor the pathogens, and the use of clean or treated seed are other important cultural practices. In Florida, Luke et al. (1983) found that septoria glume blotch was not reduced in winter wheat after two oat crops when infested wheat seed was used. However, an acceptable level of control was obtained when clean seed or benomyl-treated seed was used in conjunction with the 3-year rotation. However, the survival of inoculum is greatly influenced by the environment. Shipton et al. (1971) in a review of the common septoria diseases of wheat indicated that *Septoria* spp. in infested straw on the soil surface may infect host seedlings even two seasons later. Thus rotations of 3 years or longer may be required to curtail these diseases.

Crop rotation is a primary control method for a number of soilborne diseases, including take-all of wheat, a disease caused by *Gaeumannomyces graminis* var. *tritici* Walker. In Saskatchewan, Russell (1934) found that when wheat followed oats or fallow the incidence of this disease was greatly reduced. He recommended for its control a 2-year rotation in which wheat alternated with oats or other nonhost crops such as corn, sunflower, flax, peas, sweetclover, and potatoes, or with fallow. He noted that barley was less affected than wheat and that many cultivated and native grasses were hosts of the fungus. The disease is favored by high soil moisture and deficiencies of phosphorus and potassium; in some regions it is a more severe problem in wheat on new or recently broken land than on older cropped land (Yarham 1981). The notable occurrence of take-all in the parkland regions of the Canadian prairies during the first half of this century probably was related to some of these factors. For some years it declined, perhaps because of the increased use of fertilizer and crop rotation. In recent times, resurgence of the disease may be associated with more intensification in crop production.

In some regions, take-all is sometimes reduced under prolonged monoculture of wheat and barley. For example, in some soils the disease increases during the first several years and thereafter diminishes to a modest level. This phenomenon, called "take-all decline," appears to result from some kind of naturally occurring biological control. Cook (1981), in northwestern United States, found that, whereas three consecutive years of alfalfa, oats, or potatoes (but not soybeans or a grass mixture) provided good control of take-all in the subsequent wheat crop, all the break crops except the grass mixture nullified "take-all decline." Investigations on the occurrence of "take-all decline" in Canada have not been reported.

The preferred break crops in a rotation to reduce a disease are the nonhosts of the pathogen. Although some crops may be less affected by the pathogen than others, they may maintain it to the detriment of the next fully susceptible crop. For instance, barley typically incurs less take-all than wheat but wheat following barley may be severely diseased. Further, a number of crops may be quite effective as breaks for one disease but may differ in effectiveness for other diseases. For example, corn may be used in rotation with wheat to reduce take-all; however, in areas where Fusarium head blight or scab (*Fusarium* spp.) occurs on wheat the use of corn may exacerbate the problem (Teich and Nelson 1984).

Whereas take-all is strictly soilborne, common root rot of cereals (caused by several fungi) is both seed- and soil-borne. However, in the Canadian prairies most infections in postseedling plants are initiated by inoculum of the dominant pathogen (*Cochliobolus sativus* (Ito & Kurib.) Drechs. ex Dastur) in the soil. Chinn and Ledingham (1958) reported that 40% of the conidia of this fungus were still viable after almost 2 years in the soil. Short rotations are unlikely, therefore, to reduce this disease markedly and studies reported in the section on experiments support this concept.

The foregoing examples indicate that crop rotation can play an important role in disease management. Although advances in other methods of disease control such as improved fungicides and nematocides and cultivar resistance may lessen dependence on rotations in various situations, rotations warrant serious consideration for use as a preventative disease measure. It is axiomatic, however, that crop rotation for disease control must be compatible with the other agronomic factors within cropping systems.

Insects

Crop rotations, particularly the practice of summer-fallowing, have played an important role in reducing infestations of insect pests in western Canada. However, because most insect problems are sporadic in occurrence, obtaining data from planned experiments is difficult. Consequently, little research has been done on the effects of crop rotations on pest abundance and damage. Much available information is based on anecdotal evidence and speculation derived from our knowledge of the life history and habits of the various pests.

Rotations that include a fallow year usually greatly reduce the likelihood of insect damage, particularly to the first subsequent crop, by interrupting the buildup of resident, in-field populations. The effect is usually greatest for those pests that have limited dispersal capabilities. Rotation of different crop types such as cereals and oilseeds also reduces populations of most pest species. Conversely, recropping to the same crop permits populations to build up and increases the risk of pests reaching damaging levels. Recropping directly into standing stubble is a practice that is liable to result in an increased incidence of insect damage.

Because the crop rotations commonly used in western Canada do not differ greatly, or consistently, in their effect on the likelihood of pest damage, little effort has been devoted to designing cropping sequences that will minimize the buildup of pest populations. This situation will likely remain as long as summer fallow every 2nd or 3rd year remains a common practice.

Reduced tillage or no-till summer fallow may not be quite as effective in reducing pest abundance as conventionally tilled summer fallow, which exposes the inactive stages (i.e., eggs and pupae) to predation and desiccation. However, in practice the differences of the various types of summer fallow on insect population levels are unlikely to be sufficient to influence the choice among alternative methods of fallowing. Conservation tillage, and particularly the accompanying increase in the amount of plant residues on the soil surface, may alter the interactions between pest and beneficial arthropods as well as affecting pesticide efficacy and persistence. Whether these effects can be influenced by crop rotation schemes is not yet known. Experience elsewhere has shown that reduced or no-tillage, particularly in conjunction with continuous cropping, often results in an increased incidence of insect damage.

Some attributes of the major insect pests to be considered when planning rotations are discussed for specific groups.

Grasshoppers

Hoppers seldom lay eggs on cultivated summer fallow, even when there is heavy trash cover. Control of green growth on no-till summer fallow in late summer and on stubble after harvest will reduce egg-laying. Crops planted on stubble fields heavily infested with egg pods are likely to require treatment throughout. A fall tillage for the sole purpose of destroying grasshopper eggs is of limited value. Destruction of all green growth before eggs begin to hatch, in fields to be summer-fallowed, will starve most newly hatched grasshoppers.

Cutworms

Several kinds of cutworms hatch in the fall and overwinter as partly grown larvae. The larvae feed on weeds and volunteer crops until freeze-up and again in the spring. They usually finish feeding before spring-seeded crops emerge. Control of winter annuals and volunteer crops in the fall reduces the abundance of cutworms considerably. One species, the army cutworm (*Euxoa auxiliaris* (Grote)) sometimes causes damage to winter wheat and fall rye in southern Alberta and southwestern Saskatchewan. In years when it is very abundant this cutworm may also seriously damage spring-seeded crops in fields where sufficient green growth in fall provides food for the young larvae, but growth in the spring is insufficient to enable the cutworms to mature before the spring crop emerges.

Redbacked (*E. ochrogaster* (Guenée)) and pale western cutworm (*Agrotis orthogonia* Morrison) moths also lay their eggs in the fall, from mid August to mid September, but the eggs do not hatch until the following spring. Moths of the redbacked cutworm, which is primarily a pest in parkland areas, usually concentrate their eggs in patches of weeds on summer fallow or in crops. Control of fall weed growth reduces the likelihood of infestation by this cutworm. However, pale western cutworm moths lay their eggs in loose soil regardless of vegetation. Presence of a surface crust during the period when the moths are laying eggs discourages egg-laying by this species. However, control of heavy weed growth on tilled summer fallow should not be sacrificed in an attempt to preserve a surface crust because loss of moisture from a weedy summer fallow is a certainty, whereas infestation by cutworms is usually a low risk. Infestations of pale western cutworm are more likely to occur with longer-term rotations, especially continuous cropping, because harvest often coincides with the period of egg-laying, and harvesting operations frequently produce the pulverized soil that favors egg-laying.

Wireworms

Wireworms are long-lived insects that require 3–5 years to develop from egg to adult. A field may be infested for years before damage becomes evident. The damage is usually greatest in crops seeded on summer fallow, presumably because the wireworms are ravenous. The most important pest species are *Tenicera destructor* (Brown), the prairie grain wireworm, and *Hypolithus bicolor* Esch., (no common name). Spring wheat and spring rye are most susceptible to damage; oats, barley, canola, and flax are less likely to be damaged unless the infestation is severe. Winter wheat and fall rye are rarely damaged. Treatment of seed with insecticide gives effective and economical control; best results are obtained when the treatment is used on spring wheat following summer fallow. Indications are that populations of wireworms increase in fields that are continuously cropped and if such fields are summer fallowed the danger of damage to the subsequent crop increases.

Wheat stem sawfly

This major pest of hard red spring wheat existed on much of the Canadian prairies prior to the introduction of resistant solid-stemmed varieties. A resurgence of wheat stem sawfly (*Cephus cinctus* Norton) in some areas of Alberta and Saskatchewan in recent years has resulted from dry climatic conditions favorable to this sawfly and from a reluctance to grow the resistant varieties because of their lower yield. Recently wheat stem sawfly damage to winter wheat in southern Alberta has been reported, and in parts of Montana it has also become an important pest of winter wheat. Agronomic practices that favor an increase of wheat stem sawfly are minimum or no-tillage, strip farming, and recropping, particularly continuous cropping of susceptible varieties. If susceptible varieties are seeded on or alongside infested stubble, damage is almost certain to result. Rotation to nonsusceptible crops (i.e., resistant varieties of bread wheat, barley, oats, or oilseeds) or summer fallow will reduce sawfly populations considerably.

Wheat midge

Populations of the wheat midge (*Sitodiplosis mosellana* (Gehin)) occasionally reach high levels in areas of high rainfall, particularly after a series of years with warm, humid conditions during the egg-laying period, i.e., when the wheat is in blossom. A serious outbreak of this orange midge occurred in northeastern Saskatchewan from 1983 to 1985. The adult midge is not a strong flier and most infestations result from a progressive, within-field, population buildup promoted by continuous cropping of spring wheat.

Russian Wheat Aphid

Potentially a serious pest of wheat and barley, the Russian wheat aphid (*Diuraphis noxia* (Mord.)) was first found in North America in 1980, when it became established in central Mexico. Since then it has spread rapidly northward reaching southern Alberta and southwestern Saskatchewan in late summer of 1988. A key factor in determining whether or not this aphid becomes a persistent pest in western Canada will be its ability, or inability, to overwinter on the northern Great Plains. If it cannot overwinter here or in adjacent states of the United States, infestations will likely be sporadic, as are those of other cereal aphids such as greenbug (*Schizaphis graminum* (*Rondani*)), corn leaf aphid (*Rhopalosiphum maidis* (Fitch)), and English grain aphid (*Macrosiphum avenae* (Fabricius)), which are migratory and therefore unaffected by crop rotation. In western Canada usually only late-seeded crops are subject to damage by aphids or the diseases that they transmit. If the Russian wheat aphid is able to overwinter here, control of volunteer grains and grassy weeds in the fall will likely become more important.

Economics

Farm managers choose cropping systems or rotations and organize their resources (e.g., land, labor, machinery, capital, and management) so that economic returns of the farm are maximized. Factors to consider when making these decisions include the following:

- potential gross revenue
- costs of resources and services used in production
- level of risk involved
- long-term effects on soil productivity and economic returns.

Most producers are assumed to be interested in earning the highest possible net return (i.e., the difference between gross sales and production costs). In situations where factors of production are fixed (e.g., land), this amounts to maximizing the return to the fixed factors. In most cases, producers formulate expectations about the possible prices and marketing opportunities for products prior to crop establishment. Not only do prices have direct bearing on the level of expected net return, but also the relationships among the product prices (i.e., the relative price ratios), together with the technical rates at which products may be substituted for one another in production, are important in determining the most profitable combinations and levels of products. These price and income expectations are often influenced greatly by government policies and programs, such as grain delivery quotas, crop insurance, income stabilization schemes, transportation and input subsidies, and tax considerations.

Producers differ markedly in the nature and quantity of the physical (e.g., machinery, labor, and buildings), financial (e.g., operating and investment capital), and personal (e.g., management skills) resources available to them. Further, production alternatives differ greatly in the amounts and distribution of the inputs required. In many cases these inputs and resource services may be supplied in several ways. When operating capital is nonlimiting, producers should use each resource to the point at which the value of the additional yield obtained from the last unit of input equals its unit cost. Alternatively, when operating capital is available in limited supply, inputs should be used to the point at which the additional yield obtained per dollar of input is the same for all inputs.

Risk arises from unexpected changes in yields, market considerations, or both. Production risk originates from variations in the amounts and distribution of rainfall, temperature, insects, diseases, weeds, and other uncontrollable factors of production. Market risk arises from unexpected changes in product prices, input costs, and marketing opportunities. Producers generally attach a positive value to cropping systems that exhibit low risk or low variability. The amount of risk that producers are willing or able to accept depends on their personal preferences, past experiences, attitudes towards risk, and financial position.

Summer-fallowing is one means of reducing yield risk in the drier regions of western Canada. Summer-fallowing, by increasing soil moisture reserves, reduces the dependency of yields on growing season rainfall. Thus the year-to-year variation in yields may be lower when crops are grown on summer fallow than on stubble. Crop diversification is also considered as a means of reducing market risk. The argument is made that with a greater diversity of products to sell, there is less income dependence on short-run happenings in any single market. Crop diversification is generally most effective in reducing market risk when the correlation of prices among products is weakest (i.e., for products that are not close substitutes or those that are sold in different markets). However, there may be some additional supply risk when growing an array of crops because of the need for greater managerial expertise and differences among crops in their response to variation in growing conditions. Thus benefits of crop diversification must be determined by weighing both the market and supply risk components against those from crop specialization. Benefits from crop specialization generally arise from economies of scale and gains from experience.

Choice of a production system can greatly influence rates of soil degradation and consequently the future earnings of a farm. Frequent summer-fallowing and poor soil management practices have been implicated in excessive soil degradation (Rennie and Ellis 1978). Soil degradation generally results in two types of costs. First, on-farm or private costs refer to the value of future potential productivity that is lost. Second, social costs are related in part to the agricultural output that is permanently lost and in part to external factors such as water and air pollution. Although these costs are difficult to quantify because of the many intangibles and the fact that their effects may only be realized many years into the future, their consideration should be included (even qualitatively) in decisions on the choice of a crop rotation and on the use of production technologies aimed at soil conservation.

Legumes, pulse, and forages

Legumes and forage grasses have been an important part of crop rotations on farms in the Black and Gray soil zones (parkland) for many years. The crops are grown for seed, feed, dehydrated feed products, and green manure. On livestock farms, alfalfa, clovers, and perennial grasses are grown mainly for feed.

Many early rotation studies were designed to determine the effect of including these crops on a regular basis in cropping systems on each soil type. Ripley (1969) reported that legume and forage grasses grown in rotation with cereal crops were extremely beneficial for crop production, erosion control, and soil improvement in the Black and Gray soil zones. They were also beneficial for erosion control in the

Brown soils, but here greater than normal precipitation was essential for their use.

In areas where legumes grow well, studies (Rice and Hoyt 1980, Bowren et al. 1986, Biederbeck 1988) have shown that, if adequately inoculated, legumes will produce high yields of good-quality forage. At the same time they fix large quantities of atmospheric nitrogen, help to control soil erosion, and have other beneficial effects on the productive capacity of the soil for future crops.

At Beaverlodge, in the Peace River region of Alberta, Rice and Hoyt (1980) reported that N was fixed annually by several legume species at up to 220 $kg{\cdot}ha^{-1}$. Nitrogen fixation was lower on Black soils where the organic matter was higher than on Gray Luvisols. The amount of N added to the soil was greatly increased if the legumes were incorporated as green manure. The resulting increase in yield of subsequent crops was substantial in most years. Deep-rooted legumes are believed to improve soil permeability and aeration.

When sweetclover, alfalfa, and red clover were used for green manure on a Black soil at Melfort and a degraded soil at White Fox, Sask., they returned N to the soil at 60–95 $kg{\cdot}ha^{-1}$ in the second year (Bowren et al. 1969). However, if a crop of hay or silage was removed and the stubble worked in, then N was returned to the soil at only 10–20 $kg{\cdot}ha^{-1}$.

On soils where drought is not usually severe, pulse crops for seed have also gained in importance. They fix N and improve production while providing a cash crop (Bowren et al. 1986). In Manitoba, pulse crops fixed between 30 and 90% of their total requirements for N (L.D. Bailey, personal communication). For example, fababeans fixed up to 90% whereas peas and lentils fixed only 30–40% of the N that they need. Forage legumes, on the other hand, can supply by fixation all the N they require. Harvesting of seed or forage, however, means that only a small proportion of the fixed N is returned to the soil. Nonetheless, the harvested crop is usually worth considerably more than the cost of replacing the N with commercial fertilizer, because other unidentified soil benefits, not directly related to fixed nitrogen, will also occur (Wright and Coxworth 1987).

Recently, introduction of annual legumes adapted to the Brown soil zone has encouraged interest in their use for soil improvement through green manuring. On these soils, yields of perennial legumes are low and they fix only a small amount of N in the year of establishment. If grown for more than 1 year, their deep roots usually reduce soil moisture to such a level that yields of subsequent grain crops are severely depressed for 1–2 years (Brown 1964). In growing biennial sweetclover the customary underseeding of a cereal crop severely restricts the choice of herbicides that can be used in the cereal crop. Furthermore, the sweetclover competes with the cereal crop for moisture and nutrients; if it is not turned under early in the 2nd year its extensive root system dries out the rooting zone. In addition, cost of seed and the difficulty of obtaining a good stand have discouraged many producers in the Brown and Dark Brown soil zones from using sweetclover in crop rotations. Consequently, alfalfa and sweetclover are *not* recommended for use with wheat in short rotations on these soils.

In an effort to develop a system to adapt legumes for green manure on the drier prairie soils, several new concepts using annual legumes in the summer-fallow year are being studied (Slinkard et al. 1987, Biederbeck 1988). To be effective for this purpose the annual legume should exhibit the following characteristics:

- fast emergence to provide an early ground cover for soil protection
- high efficiency in use of water
- high rate of dry matter production and N-fixation capacity
- good quality for emergency feed supply
- good competitiveness with weeds
- good seed-yielding capacity.

Preliminary information shows that several annual legumes meet these requirements. However, the most efficient and best method for their management has yet to be determined (Biederbeck 1988).

The development of suitable crop rotations that include pulse crops is crucial for dealing with disease and weed control and for maintaining adequate residue cover to control erosion. Pulse crops have proven useful for reducing the frequency of summer fallow in the Dark Brown, Black, and Gray Luvisolic soils.

Studies have already shown that legume crops have a place in crop rotations for soil improvement throughout the prairies. If managed properly legumes reduce fertilizer inputs, improve productivity, and therefore play a significant role in the choice of crops for rotations in all soil zones.

CROP ROTATION EXPERIMENTS

MATERIALS AND METHODS

Agriculture Canada research stations in western Canada have been experimenting with crop rotations for more than 100 years (Table 3). Some quite thorough studies (e.g., at Swift Current) have determined the influence of rotation length, crop sequence, summer-fallow substitute crops, and N and P fertilizers on crop yields, grain protein, N and P uptake by the plants, moisture conservation and use, energy efficiency, economic returns, and changes in the chemical, physical, and biological properties of the soil. However, most studies have focused only on the effects on crop yield and moisture use efficiency. Studies located on research stations were usually set out on plots in a replicated and randomized block design; those at off-station sites were often set out as unreplicated tests on a field scale. All stages of each rotation were present each year and each rotation was cycled on its assigned plots. For example, two plots per replicate were assigned to each 2-year F–W[1] rotation, three plots per replicate to each of the 3-year rotations, and one plot per crop type for each continuous crop rotation.

Cultivation and tillage were performed with field-sized equipment using the generally recommended methods and practices for the region. Seedbeds were prepared on all plots to be planted, usually involving one or two operations with a heavy-duty cultivator or rodweeder (or both) or whatever implement was commonly used for a particular soil texture in a particular soil zone. Spring-seeded crops were usually planted in early to mid May at the recommended seeding rates using commercially available disc or hoe press drills; fall-seeded crops such as rye were planted during the first week of September. Forage crops and legumes for green manure were usually established by underseeding them together with the preceding cereal crop. Recommended varieties of seed were used each year, but cultivars changed over the years as new varieties became available.

Rates of application of N and P to rotations receiving fertilizer varied over the years in most studies. During the early years of most experiments (up to the early 1970s), N as ammonium nitrate–phosphorus (23–23–0) and P as monoammonium phosphate (11–48–0) were applied at relatively low rates to stubble- and fallow-seeded crops, respectively, in accordance with the rotation specifications and the generally recommended rates for fertilizers as provided for each region by the Advisory Council on Soils in each province. Since the mid 1970s, N (ammonium nitrate, 34–0–0) and P (monoammonium phosphate, 11–48–0) were usually applied at higher rates based on soil tests of the individual plots. In all cases, P was banded with the seed;

---

[1] Abbreviations are listed on pages vii and viii.

N was also banded with the seed in early years, but in later years it was broadcast and soil-incorporated either at time of planting or in the previous fall (Black and Gray soil zones only).

Table 3 Summary of long-term rotation studies undertaken by Agriculture Canada in the Prairie Provinces

	Number of studies	
Location	Completed	Continuing
Brown soil zone		
Swift Current	6	2
Indian Head	1	–
Subtotal	7	2
Dark Brown soil zone		
Lethbridge	5	2
Scott	3	4
Regina	2	–
Indian Head	6	–
Swift Current	1	–
Subtotal	17	6
Black soil zone		
Beaverlodge	2	2
Lacombe	3	0
Vegreville	1	2
Melfort	9	1
Indian Head	7	2
Brandon	9	1
Subtotal	31	8
Gray and Dark Gray soil zones		
Beaverlodge	5	2
Melfort	5	1
Scott	3	1
Subtotal	13	4
Total	68	20

Herbicides were applied as required for control of weeds in crops, using recommended types, methods, and rates. Beginning in the late 1940s, 2,4-D and MCPA formulations were used to control broadleaf weeds; barban was used at several locations to control wild oats. During the mid 1960s triallate, applied in spring or fall and incorporated into the soil with a disc or cultivator, was used to control wild oats in wheat. In recent years bromoxynil plus MCPA (1:1) was the main herbicide for controlling broadleaf weeds, and diclofop-methyl replaced triallate as the main herbicide to control wild oats and green foxtail. Glyphosate was applied prior to planting or following harvest as a spot treatment to the continuous-type rotations to control Canada thistle and quack grass in several years. Herbicide was rarely applied to the forage plots. Malathion and other insecticides were used periodically to control weevil in sweetclover and to control grasshoppers.

Annual crops were swathed at the full-ripe stage (usually late August to late September); yields were determined by threshing the grain from a given area or by using a conventional combine for the entire plot. The straw was generally redistributed onto the plots by a straw spreader attachment on the combine. Forage plots were cut at full bloom (usually mid June), field dried, baled, and the hay weighed. Crops for green manure were incorporated into the soil by discing or plowing in mid June. In the Brown and Dark Brown soil zones, cropped plots generally received 2,4-D applied in fall for control of winter annual weeds; in the Black and Gray soil zones, plots that were sown to cereals or oilseeds were usually tilled once in the fall with a cultivator or disc.

On summer-fallow areas, weeds were controlled by tillage. An average of three to six operations with a heavy-duty cultivator were required with the number increasing from Brown to Black and Gray soil zones. At several locations, a rodweeder or Noble blade cultivator replaced one or more cultivation operations in some years. In the Brown and Dark Brown soil zones, summer-fallow areas also received an application of 2,4-D in late fall to control winter annual weeds.

CROP PRODUCTION AND QUALITY

Each year producers decide which types and amounts of crops to produce and the levels and manner in which the controllable inputs are to be combined. These decisions require an understanding of how crop yields are influenced by interactions of soil, plant, and climate and by the management system to be used.

Whether advantages exist in combining crops into one rotation depends partly upon the indirect effects that each crop has on the other. Crops are complementary when an increase in the output of one crop from a given land area also leads to greater output of another. Complementary relationships usually exist only over a limited range because of the law of diminishing returns. Furthermore, the complementary effects may be masked, reduced, or eliminated by use of substitutes such as N fertilizer, pesticides, or snow-trapping techniques. These relationships eventually give way to one of competition, wherein an increase in output of one crop from a given land area is at the expense of output of another. This competitive relationship occurs either when crops have similar requirements for moisture, nutrients, and other management inputs and are grown in about the same season, or where allelopathic effects or toxins from one reduces yield in the following crop.

This section examines the influence of rotation length, crop sequence, summer-fallow substitute crops, and fertilizers on yields and quality of grains, straw, and forages.

Brown soil zone

At Swift Current, the yields of wheat grown on adequately fertilized fallow after 12 years (Campbell et al. 1983*b*) and after 18 years (Zentner and Campbell 1988) were similar whether grown in a 2-year F–W rotation or a 3-year F–W–W rotation (Table 4). Similarly, yields of wheat grown on adequately fertilized stubble were unaffected by rotation length or by the preceding crop, except when wheat was rotated continuously with flax. The yield of wheat following flax in the Flx–W–W rotation averaged about 13% less than the yield of wheat following wheat, rye, oat hay, or even flax when the rotation included fallow. The yield depression was attributed to the reduced trash cover from flax and higher infestations of weeds in the Flx–W–W rotation. There was no evidence of an allelopathic response of wheat grown on wheat stubble, nor was there evidence of complementary effects among wheat and other crops grown on stubble.

Within monocultures of wheat rotated with fallow and receiving N and P, the yield of wheat grown on stubble averaged 71–74% of that on fallow. Here, as in all other studies, annual variability of yield (Y_{cv}) was greater for wheat grown on stubble than for wheat grown on fallow, reflecting the higher stored soil moisture in fallow. Wheat yields generally increase with time (Zentner and Campbell 1988).

Alternatively, when annual grain yields were expressed on a total farm basis ($kg \cdot ha^{-1}$), the quantity of wheat produced in monoculture wheat rotations was inversely related to the proportion of fallow in the rotation (Table 5).

Grain yields (Y_g) were related to straw yields (Y_s) for several rotations (Campbell et al. 1988*b*) using 18 years of data by the relationship: $Y_s = 52 + 1.63\, Y_g$ ($r = 0.83^{**}$). Note that straw was cut at ground level.

** Significant at $P > 0.01$.

Table 4 Mean yield, protein concentration, phosphorus concentration, and volume weight of wheat by rotation-year at Swift Current, Sask., (1967–1984)

Rotation sequence†	Fertilizer††		Grain yield			Grain protein‡		P content of grain		Volume weight of grain	
	N	P	18-yr mean (kg·ha−1)	% of control	Y_{cv}§ (%)	18-yr mean (%)	% of control	18-yr mean (%)	% of control	18-yr mean (kg·hL−1)	% of control
Wheat grown on fallow‡‡											
F–**W** (control)	√	√	1898	100	30	16.2	100	0.31	100	76.3	100
F–**W**–W	√	√	1912	101	29	15.8	98	0.32	103	76.0	100
F–**W**–W	0	√	1872	99	31	16.0	99	0.32	103	76.2	100
F–**W**–W	√	0	1715	90	30	16.0	99	0.33	106	76.1	100
$S_{\bar{x}}$§§			37			0.12		0.004		0.10	
Wheat grown on stubble											
F–W–**W**	√	√	1403	74	45	15.8	98	0.40	129	75.4	99
F–W–**W**	0	√	1307	69	43	15.3	94	0.38	123	75.9	99
F–W–**W**	√	0	1263	67	45	15.8	98	0.39	126	75.3	99
F–Ry–**W**	√	√	1425	75	51	15.3	94	0.39	126	75.5	99
F–Flx–**W**	√	√	1376	73	44	16.4	101	0.37	119	75.8	99
Oat(hay)–**W**–W	√	√	1349	71	42	15.7	97	0.37	119	75.6	99

(continued)

Table 4 Mean yield, protein concentration, phosphorus concentration, and volume weight of wheat by rotation-year at Swift Current, Sask., (1967–1984) *(concluded)*

Rotation sequence†	Fertilizer†† N	Fertilizer†† P	Grain yield: 18-yr mean (kg·ha−1)	Grain yield: % of control	Grain yield: Y_{cv}§ (%)	Grain protein‡: 18-yr mean (%)	Grain protein‡: % of control	P content of grain: 18-yr mean (%)	P content of grain: % of control	Volume weight of grain: 18-yr mean (kg·hL−1)	Volume weight of grain: % of control
Oat(hay)–W–**W**	√	√	1332	70	44	15.7	97	0.36	116	75.4	99
Flx–**W**–W	√	√	1201	63	42	16.0	99	0.33	106	75.4	99
Flx–W–**W**	√	√	1271	67	44	15.7	97	0.36	116	75.2	99
Contin. **W**	√	√	1354	71	37	15.7	97	0.39	126	75.6	99
Contin. **W**	0	√	1162	61	39	14.1	87	0.40	129	76.5	100
$S_{\bar{x}}$			30			0.12		0.008		0.12	

Source: Zentner and Campbell 1988.

√ Denotes fertilizer nutrient was applied; 0 denotes fertilizer nutrient was not applied.

† Results from a W–lentil and chemical fallow–WW–WW rotations are not shown because of the few years of available data.

†† The 18-year mean rate of N application (unless otherwise indicated) was 4–7 kg·ha^{-1} for wheat grown on fallow; 22–28 kg·ha^{-1} for wheat grown on stubble in the 3-year rotations, and 25–33 kg·ha^{-1} for continuous-type rotations. P was applied as P_2O_5 at an annual rate of 22 kg·ha^{-1}.

‡ Protein = % N × 5.7.

‡‡ The boldface rotation-year denotes the one represented by the values shown.

§ Calculated over years for each treatment.

§§ $S_{\bar{x}}$ = standard error of the mean in this and subsequent tables and figures.

Table 5 Total annual wheat production for rotations fertilized with N and P

Location	Average annual production (kg·ha^{-1})				
	F–W	F–W–W	% of F–W	Contin. W	% of F–W
Swift Current (1967–1984)	949	1105	117	1354	143
Scott (1963–1978)	1018	1077	106	1330	131
Scott (1966–1984)	1265	1503	119	1675	132
Lethbridge (1972–1984)	1516	1521	100	1747	115
Indian Head (1960–1984)	1230	1464	119	1810	147
Melfort (1960–1984)	1392	1730	124	1808	130

Sources: Adapted from Kirkland and Keys 1981, Janzen 1986, Zentner et al. 1986, and Zentner and Campbell 1988.

The protein concentration (%N × 5.7) of wheat grown on fallow in the 2-year F–W rotation was greater than for wheat grown on fallow or stubble in the fertilized F–W–W and continuous wheat rotations (Table 4). Inclusion of fall rye in the rotation generally reduced the protein concentration of the subsequent wheat crop as a result of greater uptake of mineral N by the rye crop (see section on "Nitrogen and phosphorus dynamics"). In contrast, the protein concentration of wheat grown on flax stubble was greater than that found in wheat grown on wheat stubble because of lower uptake of mineral N by flax. Grain protein of fallow-seeded wheat was unaffected by either N or P, but failure to apply N to stubble-seeded wheat reduced protein concentration. Nonetheless, even the lowest protein concentration (14%) would be high enough to guarantee the top grade of Canada western red spring wheat.

In contrast to grain protein, P concentration in the grain of stubble-seeded wheat was greater than that for wheat grown on fallow (Table 4) indicating a yield-induced dilution of P under fallow conditions (Campbell et al. 1984*b*). Concentration of P in wheat grown on flax stubble was low and generally similar to that for wheat grown on fallow.

Seed density of wheat was unaffected by treatments (Table 4) and volume weight was always above the base quality criterion of 74.5 kg·hL^{-1} reflecting the high quality of wheat traditionally grown in this region.

The 18-year mean yield of flax was 798 kg·ha^{-1} (Y_{cv} = 50%) when grown on fertilized fallow and 545 kg·ha^{-1} (Y_{cv} = 74%) when grown on fertilized stubble. Yields of flax were more variable than for wheat (Zentner and Campbell 1988).

Yields of fall rye averaged 1750 kg·ha^{-1} or 8% lower than wheat yields for comparable treatments. The annual variability of fall rye yields (Y_{cv} = 59%) was also about double that of wheat. The 18-year mean dry matter yield for oat hay was 3735 kg·ha^{-1} (Y_{cv} = 83%). Protein concentrations of flax, fall rye, and oat hay averaged 22.4, 12.8, and 9.0%, respectively.

Dark Brown soil zone

At Lethbridge, yields of fallow wheat in a F–W rotation averaged 11% higher than in a F–W–W rotation (Table 6) (Pittman 1977, Freyman et al. 1982, Janzen 1987*a*). This lower performance in the

Table 6 Mean yield of wheat by rotation-year at Lethbridge, Alta., 1972–1984

Rotation sequence	Fertilizer (kg·ha^{-1}) N	P	13-yr mean yield (kg·ha^{-1})	% of control	Y_{cv}§ (%)
Wheat grown on fallow					
F–**W** (control)	0	0	2775	100	32
	0	20	2802	101	30
	45	0	2722	98	31
	45	20	3031	109	30
F–**W**–W	0	0	2332	84	30
	0	20	2641	95	30
	45	0	2460	89	31
	45	20	2654	96	27
Wheat grown on stubble					
F–W–**W**	0	0	1203	43	45
	0	20	1176	42	40
	45	0	1519	54	47
	45	20	1908	69	43
Contin. **W**	0	0	1156	42	45
	0	20	1284	46	42
	45	0	1505	54	52
	45	20	1747	63	52

Source: Janzen 1986.
§ Calculated over years for each treatment.

3-year rotation contrasts with results at Swift Current and likely reflects the minimal use made of herbicides for in-crop weed control during the early years of the study. On the other hand, wheat grown on stubble yielded similarly whether in a 3-year or continuous rotation at Lethbridge (Table 6) and during 1963–1978 at Scott (Table 7). The lower yield of wheat grown on stubble in the continuous rotation, as compared to the 3-year rotation during 1972–1984 at Scott, was attributed to high infestations of weeds and low rates of application for N applied to stubble-seeded crops (Zentner et al. 1986).

Yields from stubble-seeded wheat in monoculture rotations averaged up to 80% of yields from fallow-seeded wheat at Scott but were about half of yields of fallow-seeded wheat at Lethbridge. At Scott, yields of fallow- and stubble-seeded wheat showed no trend with time and were not influenced by inclusion of canola or alfalfa hay in the rotation. In contrast, at Lethbridge, Freyman et al. (1982) credited significant positive trends observed in yields of wheat grown on unfertilized fallow and stubble over a 20-year period primarily to improved control of weeds by chemicals.

Table 7 Mean yield of fertilized wheat by rotation-year at Scott, Sask.

	1963–1978†		1966–1984††		
Rotation sequence	Mean yield ($kg \cdot ha^{-1}$)	% of control	Mean yield ($kg \cdot ha^{-1}$)	% of control	Y_{cv}§ (%)
Wheat grown on fallow					
F–W (control)	2036	100	2529	100	27
F–W–W	1868	92	2477	98	27
$S_{\bar{x}}$			24		
Wheat grown on stubble					
F–W–W	1364	67	2032	80	35
F–Can–W‡	–	–	1969	81	35
F–W–W–O–H–H	–	–	2031	80	38
Contin. W	1330	65	1675	66	43
$S_{\bar{x}}$			45		

† *Source*: Kirkland and Keys 1981. Wheat on fallow was fertilized with N at average rates of 5 $kg \cdot ha^{-1}$ plus P_2O_5 at 22 $kg \cdot ha^{-1}$; wheat grown on stubble received N at 15 $kg \cdot ha^{-1}$ plus P_2O_5 at 15 $kg \cdot ha^{-1}$.

†† *Source*: Zentner et al. 1986. Wheat on fallow was fertilized with N at average rates of 5 $kg \cdot ha^{-1}$ plus P_2O_5 at 22 $kg \cdot ha^{-1}$; wheat on stubble in 3-year rotations received N at 29 $kg \cdot ha^{-1}$ plus P_2O_5 at 12 $kg \cdot ha^{-1}$; continuous wheat received N at 32 $kg \cdot ha^{-1}$ plus P_2O_5 at 16 $kg \cdot ha^{-1}$.

‡ Mean based on 1972–1984.

§ Calculated over years for each treatment.

Fertilizers were not assessed at Scott, but at Lethbridge fertilization greatly enhanced yields of spring wheat, especially for stubble-seeded crops (Table 6).

As found in the Brown soil zone, the relative annual productivity of monoculture wheat rotations (expressed as $kg \cdot ha^{-1}$) was directly related to length of rotation at Scott (Table 5); however, at Lethbridge the pattern was less clear-cut.

Yields of canola grown on fallow at Scott were generally similar for all rotations (Table 8). Variability of yield of canola on fallow (Y_{cv} = 53%) was twice that of wheat on fallow (Y_{cv} = 27%) possibly because seed weight of canola is susceptible to heat stress at flowering and also because of periodic infestations of stinkweed in this crop. Hay yields (data not shown) were highly variable, reflecting problems with establishing stands and with low rainfall during the growing season.

Table 8 Mean yield of canola fertilized with N and P in crop rotations at Scott, Sask., 1972–1978

Rotation sequence	7-yr mean yield ($kg \cdot ha^{-1}$)	% of control	Y_{cv}§ (%)
F–**Can** (control)	1108	100	53
F–**Can**–W	991	89	54
F–**Can**–O	1077	97	52
F–**Can**–O–H	1066	96	52
F–**Can**–W–O–H–H	1034	93	53
$S_{\bar{x}}$		25	

Source: Zentner et al. 1986.
§ Calculated over years for each treatment.

At Lethbridge, yields of grain sorghum were strongly affected by the preceding crop, being highest after fallow, next highest after wheat then continuous sorghum and lowest after barley (Janzen et al. 1987). The relatively low yields observed in the continuous sorghum and sorghum–barley rotations resulted from infestations of Russian thistle and kochia (Janzen et al. 1987).

Black soil zone

At Indian Head (Table 9) yields of wheat grown on fallow were similar for the F–W and F–W–W rotations where fertility treatments were similar and the straw was returned to the land. Removal of straw after harvest increased wheat yields on F–W–W rotations by 6%, probably because of better seed placement (Zentner et al. 1987*b*).

Table 9 Mean yields of wheat in rotations at Indian Head, Sask., 1960–1984

Rotation sequence	Fertilizer† N	Fertilizer† P	25-yr mean yield (kg·ha^{-1})	% of control	Y_{cv}§ (%)
Wheat grown on fallow					
F–**W** (control)	√	√	2459	100	25
F–**W**	0	0	2239	91	24
F–**W**–W	√	√	2551	104	24
F–**W**–W††	√	√	2615	106	23
F–**W**–W	0	0	2268	92	25
GM–**W**–W	0	0	2565	104	25
F–**W**–W–H (4 years)	0	0	2583	105	24
F–**W**–SC–O–H (4 yr)	0	0	2619	107	27
F–**W**–W–H–H–H	0	0	2801	114	26
$S_{\bar{x}}$			23		
Wheat grown on stubble					
F–W–**W**	√	√	1840	75	37
F–W–**W**††	√	√	1870	76	36
F–W–**W**	0	0	1103	45	49
GM–W–**W**	0	0	1491	61	46
F–W–**W**–H (4 years)	0	0	1433	58	43
F–W–**W**–H–H–H	0	0	1843	75	42
Contin. **W**	√	√	1810	74	38
Contin. **W**	0	0	1047	43	38
Flx–**W**–B	√	√	2334‡	95	41
Flx–**W**–B (high N)	√	√	2568‡	104	39
Flx–**W**–B	0	0	1338‡	53	61
$S_{\bar{x}}$			20		

Source: Zentner et al. 1987*b*.

† During 1960–1977 when fertilizer was applied using the general recommendations for the region, fertilized fallow-seeded wheat received N at 6 kg·ha^{-1} plus P_2O_5 at 27 kg·ha^{-1}. Wheat grown on stubble received N at an average of 24 kg·ha^{-1} plus P_2O_5 at 21 kg·ha^{-1}. Since 1978 when soil testing was started, fallow-seeded wheat received N at 5 kg·ha^{-1} plus P_2O_5 at 22 kg·ha^{-1}, but stubble-seeded wheat received N at 82 kg·ha^{-1} plus P_2O_5 at 25 kg·ha^{-1}. Wheat grown after flax with normal fertilizer application received N at an average of 49 kg·ha^{-1} plus P_2O_5 at 23 kg·ha^{-1}; the high N-fertilized wheat after flax received N at 72 kg·ha^{-1} plus P_2O_5 at 25 kg·ha^{-1}.

†† Straw was baled and removed.

‡ 1968–1984 average.

§ Calculated over years for each treatment.

Yields of wheat following wheat were similar within the fertilized monoculture rotations. Wheat grown on fertilized stubble yielded 71–75% of comparable fallow-seeded wheat when averaged over

25 years. During 1960–1977, when fertilizer was applied using the general recommendation for the region, yields of stubble-seeded wheat averaged only 65–67% those of fallow-seeded wheat. However, since 1978 when fertilizer was applied based on soil tests, yields were similar for fallow- and stubble-seeded wheat (Zentner et al. 1987*b*). Yields of fallow-seeded wheat showed no consistent trends over time; however, yields of fertilized stubble-seeded wheat generally increased with time, reflecting the combined effects of improved technologies and the increased rates of fertilizer used (Zentner et al. 1987*b*).

In contrast to the results obtained at Swift Current (Campbell et al. 1983*b*) and Brandon (Spratt et al. 1975), the yield of wheat in the Flx–W–B rotation averaged 20–32% higher than the yield of wheat following wheat in monoculture rotations (Table 9). Furthermore, under conditions of high N fertility in the Flx–W–B rotation, wheat yields were similar to those of wheat grown on fertilized fallow in monoculture rotations.

In the 2-year and 3-year wheat rotations, N and P increased yields of fallow-seeded wheat by 8–10% (Table 9). Fertilization resulted in an advantage for fallow-seeded wheat in two-thirds of the years. On stubble, the response of yields to fertilizers ranged from 67 to 92% depending on the previous crop and the rate of fertilizer application.

The inclusion of bromegrass–alfalfa or sweetclover hay in the rotation increased yields of wheat grown on fallow in unfertilized rotations by 14–24% and by 5–14% over yields of fertilized fallow-seeded wheat in monoculture rotations (Table 9). The highest average yield of fallow-seeded wheat was obtained in the unfertilized F–W–W–H–H–H rotation. Stubble-seeded wheat also showed a complementary relationship between the forage and subsequent wheat crop likely reflecting the increase in amount and improvement in quality of soil N by alfalfa (see section on "Soil quality").

The use of biennial sweetclover as green manure increased the yield of unfertilized wheat grown on the partial fallow by 13% and that grown on stubble by 35% (Table 9). Yields of wheat on green manure in this rotation were often similar to, and sometimes higher than, those for fertilized fallow-seeded wheat in the monoculture rotations. However, in the case of yields of stubble-seeded wheat, those from wheat in rotation with green manure were 19–26% lower than those from fertilized monoculture wheat.

Annual wheat production, when expressed on a total farm basis ($kg{\cdot}ha^{-1}$) (Table 5), responded similarly to trends observed in the Brown and Dark Brown soil zones.

Fertilizing with N and P (average N at 49 $kg{\cdot}ha^{-1}$ plus P_2O_5 at 23 $kg{\cdot}ha^{-1}$) increased flax and barley yields at Indian Head; higher than recommended rates of N increased yields even more (Table 10). Once again variability in yield for flax was considerably higher than for wheat; in several years poor establishment of flax resulted in crop failure. Variability of yield was also greater for barley than for wheat on stubble and for unfertilized than for fertilized treatments. Oat yields averaged 2891 $kg{\cdot}ha^{-1}$ (Y_{cv} = 34%).

Table 10 Mean yields of flax, barley, and bromegrass–alfalfa hay at Indian Head, Sask.

Rotation sequence	Fertilizer†		1960–1984		1968–1984	
	N	P	Hay yield, DM (kg·ha−1)	Y_{cv}§ (%)	Grain yield (kg·ha−1)	Y_{cv}§ (%)
Flx–W–B	0	0			591	67
Flx–W–B	√	√			909	62
Flx–W–B (high N)	√	√			1015	60
$S_{\bar{x}}$					16	
Flx–W–**B**	0	0			1310	74
Flx–W–**B**	√	√			2412	50
Flx–W–**B** (high N)	√	√			2743	52
$S_{\bar{x}}$					31	
F–W–W–**H** (4 years)	0	0	2840	51		
F–W–SC–O–**H** (4 years)	0	0	2180	67		
F–W–W–**H**–H–H	0	0	1224	80		
F–W–W–H–**H**–H	0	0	2798	48		
F–W–W–H–H–**H**	0	0	2938	48		
$S_{\bar{x}}$				51		

Source: Zentner et al. 1987*b*.

† Fertilizer was applied to flax and barley at an average rate of N at 48 kg·ha^{-1} plus P_2O_5 at 22 kg·ha^{-1} for the normal fertilized treatments, and at an average rate of N at 72 kg·ha^{-1} plus P_2O_5 at 22 kg·ha^{-1} for the high N treatments.

§ Calculated over years for each treatment.

Yields of bromegrass–alfalfa hay from 2nd- and 3rd-year stands were about 2.3 times that of 1st-year stands (Table 10), likely reflecting normal development of perennial crops and competition for moisture, nutrients, and shading by the companion crop in the 1st year. Stand establishment of hay crops was poor in the early years. The 25-year mean yield of dry matter in sweetclover hay was 3766 kg·ha^{-1} (Y_{cv} = 74%).

At Melfort (Zentner et al. 1986), yields of fertilized fallow-seeded wheat in F–W, F–W–W, and F–W–W–H–H–W rotations were generally similar, but the yield of wheat grown on fertilized (partial) fallow in the GM–W–W rotation averaged 5–7% lower (Table 11). This effect of green manure was opposite to that observed at Indian Head where the GM–W–W rotation was unfertilized and N fixation was likely not suppressed as it may have been at Melfort by N fertilization. Depression of wheat yields under GM–W–W rotation (at Melfort) may also have been related to increased populations of weeds as less herbicide and tillage were used on these rotations.

In contrast to the Indian Head results, yields of fertilized continuous wheat were 26% less at Melfort than yields of stubble-seeded wheat in the F–W–W rotation. This difference in yield response resulted mainly from the differences in fertilizer regimes after 1971 at Melfort; N rates were increased fivefold on fallow (no change occurred at Indian Head on fallow) and on stubble P rates were doubled (also twice the rate at Indian Head). At Melfort, yields of stubble-seeded wheat in the fully fertilized F–W–W rotation averaged about 89% of those of fallow-seeded wheat whereas fertilized continuous wheat averaged between 65–79%—the highest ratios occurring after the fertilizer rates were increased. Yields of wheat showed significant positive trends with time, reflecting the combined effects of improved production technologies, use of higher rates of fertilizers, and improved varieties.

Application of recommended rates of fertilizers increased yields of wheat at Melfort (Table 11). In the F–W–W and F–W–W–H–H–W rotations, fertilizer increased the yields of fallow-seeded wheat by an average 15% (significant about 88% of the time). Fertilizer increased the yields of stubble-seeded wheat by an average of 26%; (significant about 76% of the time). Benefits from fertilization of stubble-seeded wheat (and to a lesser extent fallow-seeded wheat) increased with time, after higher rates of fertilizer were applied in later years based on soil tests (Zentner et al. 1986). The 25-year mean yield of stubble-seeded wheat in the fertilized GM–W–W rotation was 8% lower than that in the comparable fertility treatment of the F–W–W rotation, but it was 24% higher than the corresponding yield for the continuous wheat rotation. This result contrasts with those obtained at Indian Head, although yields of stubble-seeded wheat in the green-manure rotation were similar to yields of fallow-seeded wheat at both places. On an annual basis, inclusion of sweetclover as a green manure, plus fertilization with N and P, produced higher yields of both fallow- and stubble-seeded wheat than the application of fertilizers alone in only 5 of 25 years. This general lack of yield benefit from including legume crops in the rotation likely reflects the high inherent fertility and organic matter content of the Melfort soil.

Table 11 Mean yields of wheat in rotations at Melfort, Sask., 1960–1984

Rotation sequence	Fertilizer† N	Fertilizer† P	25-yr mean yield (kg·ha^{-1})	% of control	Y_{cv}§ (%)
Wheat grown on fallow					
F–**W** (control)	√	√	2784	100	34
F–**W**–W	√	√	2752	99	34
F–**W**–W	0	0	2409	87	34
GM–**W**–W	√	√	2595	93	36
F–**W**–W–H–H–W	√	√	2838	102	34
F–**W**–W–H–H–W	0	0	2470	89	32
$S_{\bar{x}}$			29		
Wheat grown on stubble					
F–W–**W**	√	√	2437	88	40
F–W–**W**	0	0	1965	71	40
GM–W–**W**	√	√	2236	80	39
F–W–**W**–H–H–W	√	√	2259	81	33
F–W–**W**–H–H–**W**	√	√	2412	87	40
F–W–**W**–H–H–**W**	0	0	1922	69	37
F–W–**W**–H–H–**W**	0	0	2057	74	50
Contin. **W**	√	√	1808	65	47
Contin. **W**	0	0	1327	48	42
F–Can–**W**	√	√	3174‡	99	
F–Can–**W**–H–H–Can	√	√	2678‡	84	
F–Can–**W**–W	√	√	2848‡	89	
F–Can–W–**W**	√	√	2498‡	78	
$S_{\bar{x}}$			33		

Source: Zentner et al. 1986.

† During 1960–1971, when fertilizer was applied using the general recommendations for the region, fertilized fallow-seeded wheat received N at an average rate of 7 kg·ha^{-1} plus P_2O_5 at 32 kg·ha^{-1}. Fertilized stubble-seeded wheat received N at 27 kg·ha^{-1} plus P_2O_5 at 22 kg·ha^{-1}. Since 1972 when soil testing was started, fallow-seeded wheat received N at 32 kg·ha^{-1} plus P_2O_5 at 40 kg·ha^{-1} and stubble-seeded wheat received N at 70 kg·ha^{-1} plus P_2O_5 at 40 kg·ha^{-1}.

‡ 1977–1984 average.

§ Calculated over years for each treatment.

Wheat following wheat in a rotation that periodically included a forage crop yielded 6% less than wheat following bromegrass–alfalfa in the same rotation and 8% less than wheat following wheat in the F–W–W rotation with comparable fertilizers. Yields of wheat following the grass–legume crop were likely enhanced by the higher amount of readily available N supplied by the legume (see section on soil quality) and the higher level of soil moisture conserved by partial fallowing in the year of breaking the sod. Here too, total wheat production was inversely related to the proportion of summer fallow in the rotation (Table 5).

Yields of canola at Melfort decreased as rotation length increased and were greater on fallow than on stubble even after plowdown of the legume-containing forage (Table 12). This result suggests that any nutrient benefit accruing to the canola from the legume is more than offset by the forage crop using up too much moisture. The variability in yields of canola was about 25% greater than that of wheat.

Over the period 1960–1984, bromegrass–alfalfa hay yields for the fertilized rotation were 32% greater in the 2nd year of use than in the 1st year. However, based on the last 12 years, yields were unaffected by year of hay use (Table 13). The application of fertilizer increased hay yields in 8 of every 10 years, the average increase being 26–38% and the relative response being greatest for the 1st year of hay use.

A study, carried out at Brandon on Assiniboine clay loam and Miniota sandy loam soils to determine the effectiveness of summer-fallow substitute crops on wheat production (Spratt et al. 1975), showed that yields of wheat after summer fallow were not significantly higher than after sweetclover on the clay loam and no higher than after sweetclover, oat hay, corn, or potatoes on the sandy loam (Table 14). The highest concentration of wheat protein was obtained for first-crop wheat after summer fallow or sweetclover probably because of greater accumulation of nitrates in soils in these systems (see section on "Nitrogen and phosphorus dynamics").

Table 12 Effect of rotation on mean yields of canola receiving N and P at Melfort, Sask., 1972–1984

Rotation sequence	13-yr mean yield ($kg \cdot ha^{-1}$)	Y_{cv}§ (%)
F–**Can**–W	1397	43
F–**Can**–W–H–H–Can	1252	42
F–Can–W–H–H–**Can**	1137	45
F–**Can**–W–W	1171	34
$S_{\bar{x}}$		41

Source: Zentner et al. 1986.

§ Calculated over years for each treatment.

Table 13 Mean yields of bromegrass–alfalfa hay at Melfort, Sask.

	Fertilizer†		1960–1972		1973–1984	
Rotation sequence	N	P	DM yield (kg·ha−1)	Y_{cv}§ (%)	DM yield (kg·ha−1)	Y_{cv}§ (%)
F–W–W–**H**–H–W	√	√	2870	67	4090	41
F–W–W–H–**H**–W	√	√	3778	38	4161	38
F–W–W–**H**–H–W	0	0	2081	66	3001	37
F–W–W–H–**H**–W	0	0	2995	46	3319	44
$S_{\bar{x}}$			59		110	

Source: Zentner et al. 1986.

† N was applied at an average rate of 81 kg·ha^{-1}; since 1973, P as P_2O_5 was also applied at the rate of 33 kg·ha^{-1}.

§ Calculated over years for each treatment.

Table 14 Mean yield and protein content of wheat grown after summer fallow and summer-fallow substitutes on Black clay loam (CL) and sandy loam (SL) soils at Brandon, Man., 1965–1970

Wheat sequence	Grain yield ($kg \cdot ha^{-1}$)		Grain protein (%)	
	CL	SL	CL	SL
First crop wheat after:				
summer fallow	2375	1175	14.2	16.5
sweetclover hay	2337	1222	14.3	16.3
oat hay	2098	1150	13.4	15.8
corn silage	2215	1196	13.8	15.5
flax	1550	1053	13.6	15.7
potatoes	2240	1218	13.6	16.1
$S_{\bar{x}}$	38	18	0.1	0.2
Second crop wheat after:				
summer fallow	1625	1025	13.6	15.6
sweetclover hay	1472	823	13.4	15.4
oat hay	1617	1056	13.2	15.6
corn silage	1585	1081	13.3	15.7
flax	1596	988	13.1	15.4
potatoes	1552	1145	13.4	15.3
$S_{\bar{x}}$	18	24	0.1	0.2

Source: Spratt et al. 1975.

Gray and Dark Gray soil zones

The importance of the addition of both manure and fertilizers for maximizing wheat production in rotations on Gray soils was shown in a study at Beaverlodge (Table 15). Although phosphate fertilizer (11–48–0) produced substantial increases in wheat yields, increases from manure were considerably greater, and highest average yields were obtained from the combination of P and manure. Wheat yields in the GM–W–W rotation were slightly lower than in the F–W–W rotation. Inadequate supply of N likely limited yields in all rotations.

A crop rotation experiment involving sweetclover GM–W–W carried out on a Gray Luvisol at McLennan, Alta., (A.M.F. Hennig, unpublished data) showed that bromegrass–alfalfa (2 or 3 years in a 5-year rotation) was more effective for increasing wheat yields than sweetclover (Table 16). Wheat yields in the rotations with bromegrass–alfalfa were essentially equal to, or greater than, wheat

Table 15 Effects of manure and chemical fertilizer on total annual yields of wheat in rotations at Beaverlodge, Alta., 1949–1960

Rotation and fertilizer treatment†	12-year mean annual yield	
	($kg \cdot ha^{-1}$)	% of F–W–W control
F–W–W		
control	1567	100
P fertilizer	1903	121
manure	2260	144
P fertilizer + manure	2307	147
GM–W–W		
control	1426	91
P fertilizer	1789	114
manure	2031	130
P fertilizer + manure	2146	137
Contin. W		
control	1278	82
P fertilizer	1567	100
manure	2065	132
P fertilizer + manure	2199	140

Source: Ukrainetz and Brandt 1986.
† P fertilizer was applied as 11–48–0 at 34 $kg \cdot ha^{-1}$; manure at 25 $t \cdot ha^{-1}$ was applied every 3rd year.

in a F–W–W rotation. Thus the greater buildup of nitrogen from alfalfa and improvement in soil structure caused by the bromegrass in the longer rotations were the major factors contributing to higher yields. Moisture was not a limiting factor in this region.

Wheat was grown continuously at McLennan after breaking red fescue, bromegrass, alfalfa, and bromegrass–alfalfa sods that had been in production for 2 to 6 years (Hoyt and Hennig 1971). Over a 5-year period, wheat yielded more after the alfalfa-containing systems than after fallow- or grass-containing systems (Table 17). The length of time the forage stands had been in production had no bearing on their effect on succeeding wheat yields. The advantages of alfalfa and bromegrass–alfalfa over grasses and fallow were still evident in the tenth wheat crop. When fertilizer supplying N at a rate of 56 $kg \cdot ha^{-1}$ was applied to the fifth wheat crop, yields were similar for all forage treatments and were considerably higher than yields of wheat on fallow.

Table 16 Wheat yields in rotations on a Gray Luvisol at McLennan, Alta.

Rotation sequence	Yield ($kg \cdot ha^{-1}$) Unfertilized 1952–1972	% of control	Yield ($kg \cdot ha^{-1}$) Fertilized 1962–1972	% of control
Wheat grown on fallow or partial fallow				
F–**W**–W (control)	1269	100	2096	100
GM–**W**–W	1166	92	1841	88
F–**W**–W–H–H	1513	119	2845	136
Wheat grown on stubble				
F–W–**W**	976	77	1531	73
GM–W–**W**	778	61	1692	81
F–W–**W**–H–H	1101	87	1977	94
W–W–H–H–H	1629	128	2041	97
W–**W**–H–H–H	1378	108	1918	92

Source: A.M.F. Hennig unpublished data.

Table 17 Yields of wheat grown successively after a fallow-wheat rotation or forage species that had produced hay for 2–6 years on a Gray Luvisol at McLennan, Alta.

Control and preceding forage	Mean yields of wheat ($kg \cdot ha^{-1}$) Crop 1	Crop 2	Crop 4	Crop 5	Crop† 5
Control					
Fallow-seeded wheat	1320	840	640	1050	2220
Preceding forage					
Red fescue	670	690	910	1210	2580
Bromegrass	680	580	880	1300	2780
Alfalfa	2260	2530	1120	1760	2770
Bromegrass–alfalfa	1990	1160	1060	1570	2740

Source: Hoyt and Hennig 1971.

† N applied at 56 $kg \cdot ha^{-1}$.

In a similar 14-year study, carried out on a Gray Luvisol in northwestern Saskatchewan, wheat was grown after several grain crops or after alfalfa, bromegrass, bromegrass–alfalfa, and summer fallow to determine their effect on yields and quality of the subsequent wheat crop (Brandt 1981). The average yield was highest for wheat seeded on summer fallow. Yields of wheat following alfalfa were 92% of those for wheat seeded on summer fallow. The forage crops, particularly alfalfa, had a cumulative, beneficial effect on soil fertility, and yields of wheat following these crops increased with time relative to the summer-fallow crop (Fig. 3). In this study, and in contrast to others mentioned previously, the yields of wheat after canola and flax were consistently higher than after cereal grain crops. Protein concentration of wheat after alfalfa was similar to that for wheat grown on summer fallow and averaged about two percentage points higher than for wheat following grain crops.

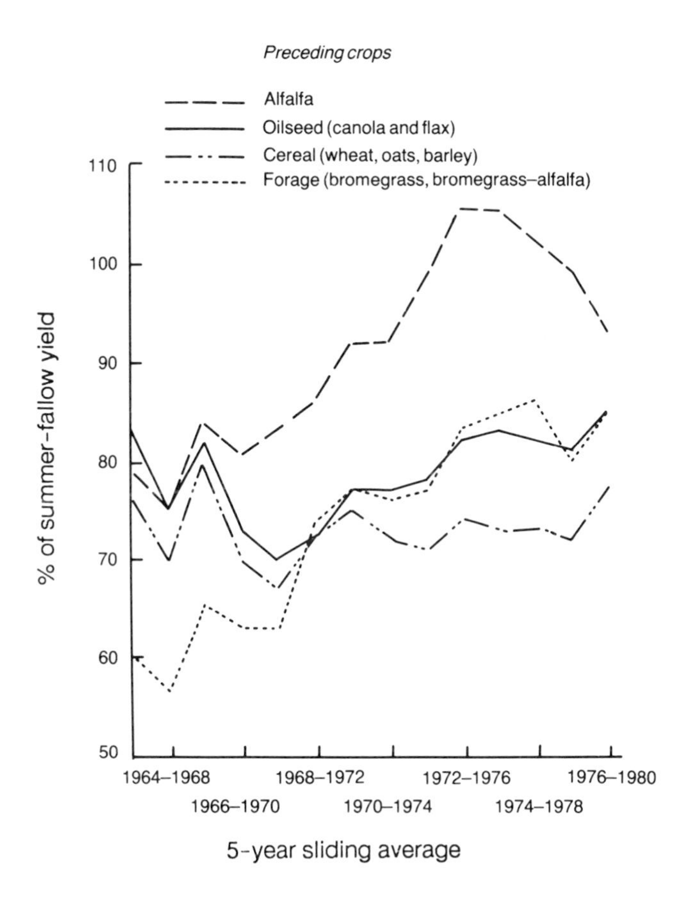

Fig. 3 Wheat yields following several crops as percentage of summer fallow yields at Loon Lake, Sask., (redrawn from Brandt 1981).

Wheat yields and grain protein in a F–W–H–H–W–W rotation were much higher than those in a F–W–W rotation on a Dark Gray soil at Somme in northeastern Saskatchewan (Table 18) (Bowren 1984). Another study on a similar soil showed that the yield of wheat following soil-incorporated legumes was equal to yield of wheat on stubble fertilized with N at 76 $kg \cdot ha^{-1}$ (Bowren and Cooke 1975).

Table 18 Comparison of two crop rotations on a Dark Gray soil at Somme, Sask.

	28-yr mean yield		
Rotation sequence	($kg \cdot ha^{-1}$)	% of control	21-yr mean grain protein conc. (%)
Wheat grown on fallow			
F–**W**–W (control)	2695	100	15.0
F–**W**–H–H–W–W	2975	110	15.6
Wheat grown on stubble			
F–W–H–H–**W**–W	2626	97	15.4
F–W–H–H–W–**W**	1994	74	15.3

Source: Bowren 1984.

Solonetzic soils

An unreplicated, long-term crop rotation was conducted on a solonetzic soil at Vegreville, Alta., during 1958–1984 (McAndrew 1986). Mean yields of these rotations up to 1980 are shown (Table 19). No statistical analyses were possible, but some general trends are apparent. Yields of spring wheat on fallow and on partial green-manure fallow were similar, but yields generally decreased with increasing number of hay crops included in rotations; continuous wheat gave lowest yields. For barley the yields from F–B rotations were highest and from continuous barley were lowest; yields generally decreased with increasing years of hay. Oat yields appeared to be highest for the 6-year mixed cereal–hay rotation. As observed at other sites, first-year hay yields were lowest, thereafter hay yields were generally similar up to the fourth hay crop, but decreased sharply in the 5th year. The results suggest that when these soils are cropped to continuous wheat and rotations including several years of bromegrass–alfalfa hay they develop poor soil structure and show moisture stress much like Dark Brown soils even though they are located in an area of much more favorable moisture.

Table 19 Mean yields of crops grown in a nonreplicated rotation study on a solonetzic soil at Vegreville, Alta., 1958–1980

	Mean annual yield		
Rotations†	Grain (kg·ha−1)	Forage, DM (kg·ha−1)	% of control
Wheat yields			
F–W (wheat control)	2286		100
F–W–B	2569		112
GM–W–B	2192		96
GM–W–O–B	2246		98
H–W–O–B–H–H	1937		85
H–W–O–B–H–H–H–H	1755		77
Contin. W	1513		66
Barley yields			
F–B (barley control)	2566		100
F–W–B	2276		89
GM–W–B	2136		83
GM–W–O–B	1872		73
H–W–O–B–H–H	2055		80
H–W–O–B–H–H–H–H	1851		72
Contin. B	1420		55
Oat yields			
GM–W–O–B	2035		–
H–W–O–B–H–H	2462		–
H–W–O–B–H–H–H–H	2222		–
Bromegrass–alfalfa hay yields			
H–W–O–B–H–H		2379	–
H–W–O–B–H–H		3494	–
H–W–O–B–H–H		3035	–
H–W–O–B–H–H–H–H		2159	–
H–W–O–B–H–H–H–H		3133	–
H–W–O–B–H–H–H–H		3069	–
H–W–O–B–H–H–H–H		3088	–
H–W–O–B–H–H–H–H		2536	–

Source: McAndrew 1986.

† From 1961–1980 fallow crops received N at 7 kg·ha^{-1} and P_2O_5 at 32 kg·ha^{-1}, stubble crops N at 24 kg·ha^{-1} and P_2O_5 at 13 kg·ha^{-1}, and hay crops N at 76 kg·ha^{-1}. From 1972 supplementary fertilizer was applied across a part of each plot to provide three subtreatments as follows: N at 76 and 152 kg·ha^{-1} and N at 157 kg·ha^{-1} plus P_2O_5 at 58 kg·ha^{-1}.

CROP PESTS

Weeds

Few of the crop rotation studies had the assessment of weed dynamics as a major goal. However, some studies recorded shifts in weed populations and herbicides used to control weed problems.

Brown soil zone

After about 5–6 years of continuous wheat in the long-term study at Swift Current, grassy weeds such as quack grass and green foxtail became a serious problem (Campbell et al. 1983*b*, Zentner and Campbell 1988). Heavy doses of expensive herbicides such as glyphosate were required to eradicate these weeds. A serious weed problem was also observed after the flax phase in the Flx–W–W system because flax is a poor competitor against weeds. The only other rotation that presented a problem with weed control was in the production of lentils on stubble. There infestations of both broadleaf and grassy weeds were high because of the crop's slow rate of growth and the limited number of herbicides that are registered for use on lentils.

Dark Brown soil zone

In a continuing study initiated in 1912 at Lethbridge in which weeds and herbicides have been observed, weed populations changed from predominantly broadleaf types, prior to the arrival of 2,4-D in the late 1940s, to wild oats in the 1950s (Freyman et al. 1982). As wild oats were brought under control with triallate in the early 1960s and other herbicides later on, green foxtail became the most bothersome weed. However, even this weed has now succumbed to diclofop-methyl so that weeds are not a major problem in this study at present.

At Scott, Brandt (1984) observed that stinkweed was less of a problem in rotations in which canola appeared every 4–6 years than in rotations including it every 2–3 years. He also reported that the green foxtail population was reduced in wheat grown in rotations where trifluralin had been used for weed control in canola.

Black soil zone

In a 6-year study carried out on a clay loam and a sandy loam in western Manitoba, Spratt et al. (1975) evaluated five intertilled and hay crops as fallow-substitutes in a 3-year rotation that included 2 years of wheat; F–W–W was used as the control. Sweetclover, potatoes, corn, oat hay, and flax were the substitute crops. Populations of wild oats increased with time on the clay loam when sweetclover or flax were present in the rotation. Green foxtail proved to be a problem on both soils but was highest in the flax system on the sandy loam. The Flx–W–W system had a major problem with quack grass and a minor problem with wild buckwheat on the clay loam soil. Quack grass was also a problem in the sweetclover system. In contrast, in the intertilled crop systems such as corn and potatoes, weeds were controlled as effectively as on fallow.

The amount of research done on the interaction of rotations and weeds has been miniscule and ad hoc compared to the known importance on this facet of agronomy.

Diseases

The influence of crop rotation on diseases has been investigated to some extent in western Canada; several studies are in progress. Information on the effects on crop production can be derived from evaluation of disease in rotation experiments, from assessing disease in the same commercial fields over time, and also from surveys in which levels of diseases are associated with histories of cropping.

In a rotation study in southeastern Saskatchewan, Ledingham (1961) found that common root rot in wheat declined as the interval between the crop increased. The average intensity of the disease over 3 years was about 27% in wheat after wheat or fallow, about 22% in wheat after a 2-year break (oats and fallow), 17% after a 3-year break (sweetclover, oats, and fallow), and 14% after a 5-year break (oats, bromegrass–alfalfa × 3, and corn). The primary cause of the disease *Cochliobolus sativus* has been reported to occur on all these break crops. However, they maintain greatly reduced levels of inoculum for subsequent infections and are considered to be poor hosts as compared to wheat or barley. In another study in west-central Saskatchewan (R.D. Tinline, personal communication), common root rot was somewhat higher in continuous wheat than in wheat in rotations with other crops. It was least in a mixed legume–cereal rotation (Table 20). The amount of *C. sativus* inoculum in the soil carrying a wheat crop differed appreciably only in the long rotation. In central Alberta, Piening and Orr (1988) assessed the incidence of common root rot in barley for 4 years following various 4-year rotation sequences involving barley, oats, canola, fallow, and bromegrass. The disease was markedly reduced both by 4 years of bromegrass and by 3 years of bromegrass with one of oats; it was slightly reduced by 2 years of bromegrass and canola after fallow. The amount of *C. sativus* inoculum in the soil was highest after barley or fallow and lowest after bromegrass; it increased with the number of successive barley crops grown.

In Saskatchewan, G.A. Petrie (personal communication) found that a rotation providing at least 3 years between canola crops usually reduces the severity of blackleg caused by *Leptosphaeria maculans* (Desm.) Ces. & de Not. He showed that the fungus can survive in stubble of previously infected plants for more than 5 years but that its viability diminished with age. The quantity of effective inoculum from 4- or 5-year-old stubble may be insufficient to generate severe disease. Most damaging infections arise from inoculum in 2- or 3-year-old stubble. Techniques for controlling the disease include the destruction of volunteer canola and wild mustard (the only other major host) during the break period, and the treatment of seed to minimize spread of the pathogen.

A long interval between susceptible crops reduces stem blight (sclerotinia stem rot) of canola. Williams and Stelfox (1980) in Alberta found that the number of viable sclerotia of the causal fungus, *Sclerotinia sclerotiorum* (Lib.) de Bary, were unchanged after three consecutive barley crops following canola. In Saskatchewan, Morrall and Dueck (1982) reported that several fields that were kept out of canola for 4 or 5 years contained abundant viable inoculum. A wide range of dicotyledonous crops are attacked by the pathogen, including safflower, sunflower, flax, peas, lentils, beans, alfalfa, clover, potatoes, and sugar beet. Cereals and grasses are nonhosts and can be used as break crops in rotations.

In a 1987 survey in Saskatchewan, W. McFadden (personal communication) observed that leaf spot (tan spot and septoria leaf blotch) prior to the flag leaf stage in winter wheat appeared less prevalent when the crop was grown after conventional fallow or canola and mustard than after chemical fallow or wheat or barley. Near plant maturity, however, the severity of the disease was fairly comparable in the fields except it was lower in wheat after canola and mustard. In the Dark Brown soil zone, Brandt and Kirkland (1986) noted that leaf diseases in continuous wheat were no more severe than in wheat grown on fallow. Although the potential does exist for increased losses from disease where continuous monoculture is practiced, the dry climatic conditions of the Brown and Dark Brown soil zones may keep such losses to a minimum.

Table 20 Incidence and intensity of common root rot in wheat grown annually or in rotation, and the amount of *C. sativus* inoculum in the upper 8 cm of soil cropped to wheat

Rotation	Incidence† (%)	Intensity† (%)	Inoculum density‡ (propagules per gram)
Contin. **W**	71	22	162
F–**W**	63	19	182
F–W–**W**	55	17	168
F–Can–**W**	56	18	153
F–Can–**W**–B–H–H	51	16	60

† 7-year average.
‡ 2-year average.

Insects

No studies are reported in which insect pests were monitored. At Swift Current fall rye seedlings were described as being eaten by grasshoppers, thus destroying or deleteriously affecting yields in 1 or 2 years. However, although these problems must have occurred and insecticides may have been used, this aspect of rotational cropping has not been regularly assessed.

SOIL MOISTURE

Moisture is the factor most limiting to crop production on the prairies. To optimize production producers are obliged to maximize the conservation and efficient use of moisture. This section examines the effect of crop rotations and fertilization on the conservation and use of moisture.

Brown soil zone

The effects of crop rotation and fertilization on moisture conservation and efficiency of moisture use (M_e) were assessed on a loam soil at Swift Current between 1967 and 1984. Six spring wheat rotations (Table 4) were examined; either they were fertilized at recommended rates of N and P after soil tests, or N or P was withheld according to rotation specifications. In this study M_e was calculated as the yield of grain divided by moisture used (i.e., precipitation from 1 May to 31 August plus moisture used from the soil between seeding and harvest). During the fall following harvest, 8–11 mm of the precipitation received was stored in soil (Campbell et al. 1987). Moisture stored during the first winter was three to five times as much as that stored during the fall. Continuous wheat fertilized with N and P stored 12 mm more moisture than that fertilized with only P. This result was credited to a larger amount of crop residues left by the treatment receiving both N and P. In the first 9 months of summer fallow, 33% of the precipitation was stored, but over the entire 21-months fallow period only 18% was stored. These results confirm earlier findings of Staple and Lehane (1952) for the 1940s in southwestern Saskatchewan.

At seeding, fallow-seeded wheat that received N and P annually had 42 mm more moisture in the 120-cm soil profile than stubble-seeded wheat that received N and P (Table 21). Failure to apply P to fallow-seeded wheat in the F–W–W rotations reduced stored moisture at seeding by about 8 mm because reduced crop growth resulted in less straw to trap snow.

Up to the shot-blade stage, moisture retention in fallow soils seeded to wheat was greater than in stubble soil seeded to wheat. By harvest, differences were not apparent (Fig. 4) and no moisture was available in the top 90 cm of soil (Campbell et al. 1987).

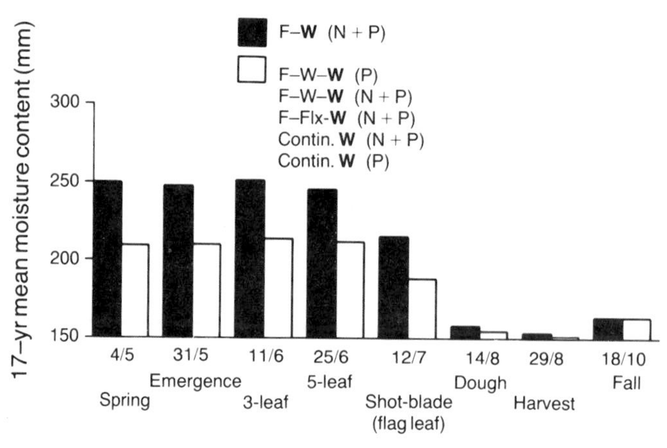

Fig. 4 Volumetric moisture content in top 120 cm of soil at various sampling times (redrawn from Campbell et al. 1987). (Data are for 1967–1984, with 1983 excluded because sampling was missed at two growth stages.)

Wheat made more efficient use of stored soil moisture in dry years than in wet years (Campbell et al. 1987). On average, wheat grown on fallow fertilized with P used 102 mm of soil moisture per year; without P, moisture used dropped to 93 mm. On stubble receiving N and P, wheat used 61 mm and continuous wheat receiving no N used 50 mm.

Table 21 18-year average and variability in available† soil moisture in spring at Swift Current, Sask., 1967–1984

Rotation sequence	Fertilizer N	Fertilizer P	Available spring soil moisture ($mm \cdot 120\ cm^{-1}$) Avg	Max	Min	Y_{cv} (%)
F–**W**	√	√	103	146	46	31
F–**W**–W	√	√	106	159	41	30
F–**W**–W	0	√	101	163	30	38
F–**W**–W	√	0	97	146	38	30
F–Flx–**W**	√	√	62	148	0	98
F–W–**W**	√	√	64	152	0	89
F–W–**W**	0	√	65	153	8	80
F–W–**W**	√	0	63	155	8	88
Contin. **W**	√	√	62	158	0	95
Contin. **W**	0	√	55	146	4	99

Source: Campbell et al. 1987.

† Moisture exceeding the lower limit of availability (Ritchie 1981) (i.e., 148 mm per 120-cm depth in this soil).

The 18-year average M_e was as high as 6.9 kg·ha^{-1}·mm^{-1} for wheat on fallow that received P and as low as 5.1 kg·ha^{-1}·mm^{-1} for continuous wheat receiving no N (Fig. 5). These values were much greater than those reported 30–40 years ago for this area (Staple and Lehane 1952, 1954) and reflect improvement in crop management and crop varieties. The improvements in M_e resulting from fertilizer was greater in later years because of the cumulative effect of fertilizer on soil quality, crop production, and thus crop residues, which enhanced moisture storage (Campbell et al. 1987). When efficiency of moisture use was based on the precipitation received from harvest to harvest, continuous wheat receiving N and P had the highest efficiency (3.75 kg·ha^{-1}·mm^{-1}) and the 2-year F–W rotation the lowest (2.60 kg·ha^{-1}·mm^{-1}), reflecting the inefficiencies of storing precipitation received during the 21-month fallow period (Campbell et al. 1987).

The effects of crop rotation and fertilization on the quantitative relationship between the yield of spring wheat (Y_g) and moisture use (M_u) were determined from the same Swift Current experiment (Campbell et al. 1988*a*). In this case M_u was defined as the sum of moisture used from the soil plus the precipitation received between 1 May and 31 July (not 31 August as used before) as these dates were found to be more precise for predictive purposes.

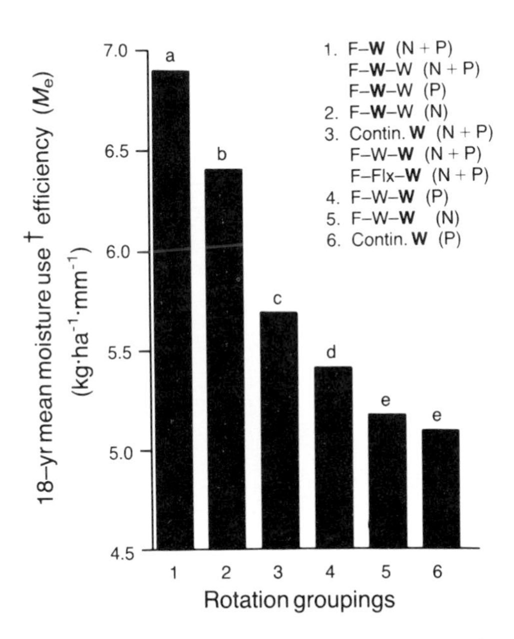

Fig. 5 Effect of rotation-year and fertilizer on annual moisture use efficiency (M_e), 1967–1984. Rotations that were not significantly different ($P > 0.05$) were grouped (redrawn from Campbell et al. 1987). (Values for groups of treatments followed by same letter are not significantly different $P > 0.05$.)

The following relationship (significant at $P < 0.001$) was derived to predict grain yield ($kg \cdot ha^{-1}$) as a function of moisture use (mm) for the well-fertilized, combined fallow- and stubble-crop system:

$$Y_g = 10.07\,(M_u - 54.02)\ (r = 0.79^{***}) \qquad [1]$$

If growing season precipitation was taken as the period 1 May to 31 August the equation was:

$$Y_g = 9.20\,(M_u - 71.85)\ (r = 0.74^{***}) \qquad [2]$$

If one uses equation 2 to estimate the amount of moisture required to produce the first kilogram of grain per hectare (i.e., the initial yield point), the value would be 72 mm and each additional millimetre would produce 9.2 $kg \cdot ha^{-1}$ of grain. The lower threshold values showed no effect of fertilizer, but the yield increase per millimetre of M_u was generally greater for the better-fertilized rotations (Campbell et al. 1988*a*). The main difference in these Y_g versus M_u relationships compared to those reported for 1925–1950 (Staple and Lehane 1954) was that the initial yield point decreased from about 140 mm to 46 mm (Fig. 6). Thus improvement in moisture use efficiency in recent years has resulted from our ability to obtain a viable crop from less available water, probably as a result of better herbicide availability, cereal varieties, and timeliness of field operations.

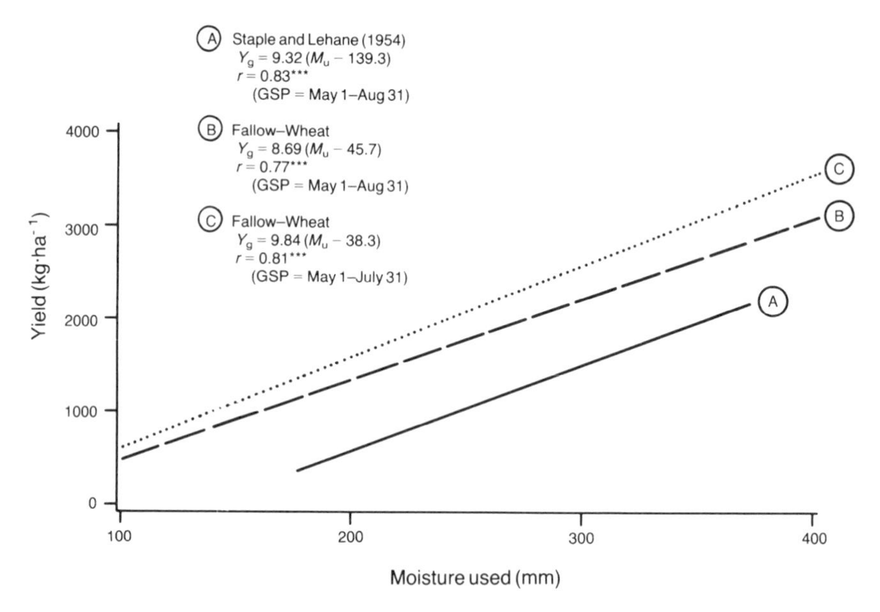

Fig. 6 Relationship between yield (Y_g) and M_u for wheat grown on fallow—our results compared to those of Staple and Lehane (1954). Data for 1970 and 1983 were omitted from our data (redrawn from Campbell et al. 1988*a*).

It was estimated that for fallow-seeded wheat, the relative effect of precipitation during the growing season (M_{gs}) on variability of yield was about 5.4 times as great as that of the available moisture in spring, whereas for stubble-seeded wheat M_{gs} was only 1.5 times as great (Campbell et al. 1988*a*). The M_{gs} was equally important in affecting yields of wheat grown on fallow or stubble.

In semi-arid regions, distribution of precipitation is almost as important as the amount of precipitation received. The grain-filling period was confirmed as the most important time for the occurrence of precipitation for both fallow- and stubble-seeded wheat. However, precipitation at or near seeding time was almost as important for stubble-seeded wheat as this ensures the establishment of an adequate stand of plants (Fig. 7).

Soil moisture data from this experiment were used to develop regression equations that may be used to relate straw yields of spring wheat and N and P content of grain and straw to M_u (Table 22) (Campbell et al. 1988*b*). Highly significant linear regressions were obtained for straw yield and N and P contents of grain versus M_u, but other relationships shown in Table 22 were less reliable (r = 0.41** to 0.55**) though still significant. These relationships can be used both when estimating approximate requirements for fertilizer N and when examining immobilization and mineralization of N in soil.

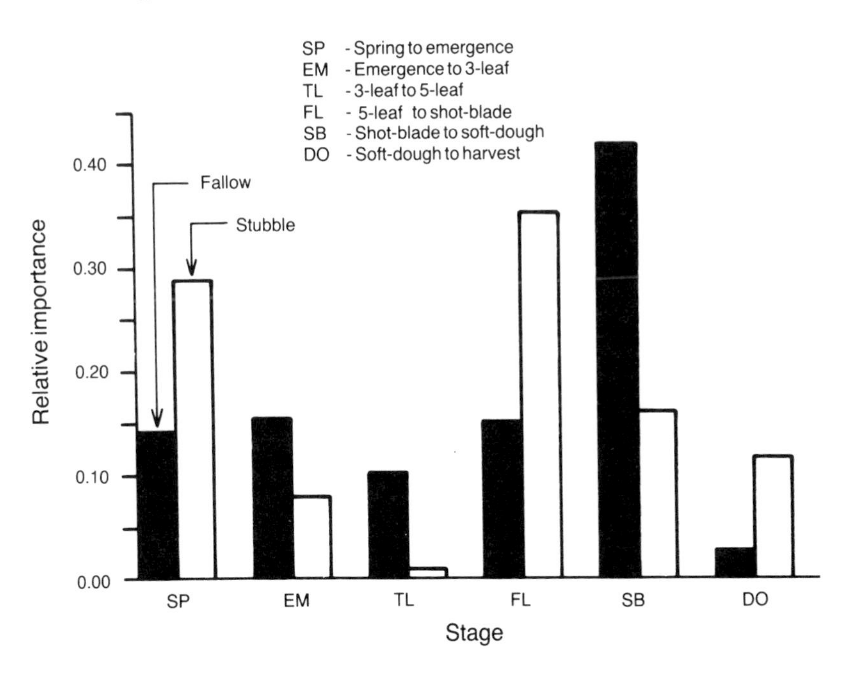

Fig. 7 Growth periods when precipitation was most essential to yields, based on 18 years of data (redrawn from Campbell et al. 1988*a*). (Calculated using normalized standard partial regression coefficients for yield versus average daily period precipitation.)

Table 22 Relationships between grain N (G_n), grain P (G_p), straw yield (Y_s), straw N (S_n), and straw P (S_p) and M_u† for well-fertilized rotation-year at Swift Current, Sask.

Treatment	Equations††	Correlation (r)
Combined fallow- and stubble-seeded systems‡	$G_N = -9.42 + 0.21\,M_u$	0.76*** (DF = 244)
	$G_P = -2.73 + 0.033\,M_u$	0.80*** (DF = 244)
	$Y_s = -1031 + 14.5\,M_u$	0.66*** (DF = 195)
	$S_N = 2.42 + 0.046\,M_u$	0.41*** (DF = 195)
	$S_P = 0.11 + 0.0029\,M_u$	0.41*** (DF = 195)

Source: Campbell et al. 1988*b*.
† $M_u = M_{sp} - M_{hv} + M_{gs}$ (1 May–31 July) in mm.
†† Units for N, P, and Y are kg·ha^{-1}.
‡ F–W, F–**W**–W, F–W–**W**, F–Flx–**W**, and Contin. **W** (N and P) for G_N and G_P, but for Y_s and S_P relationships F–**W**–W was not included.
*** Significance at $P < 0.001$; DF, degrees of freedom.

Dark Brown soil zone

In a 16-year study (1963–1978) at Scott, snow ridging, snow trap strips, row-crop fallow, standing stubble, and bare fallow were compared as methods of conserving snow and moisture (Kirkland and Keys 1981). The most effective means of trapping snow was to leave stubble standing such as in the F–W–W and continuous wheat rotations (Table 23). Snow ridging on fallow and sunflower trap strips did not significantly increase moisture reserves (from September to May) as compared to bare fallow. Widely spaced corn rows on fallow increased moisture reserves to levels comparable to that conserved by standing stubble in the two continuous wheat rotations. Standing stubble in a continuous wheat rotation conserved 50–60 mm more moisture between September and May than bare fallow.

At Scott, results were similar to those reported at Swift Current. Rotations containing a fallow treatment, with the exception of row-crop fallow, had significantly higher moisture reserves (30–50 mm) available for the crop seeded on it than was available for those crops seeded on stubble in the more extended rotations (Table 23). Further, when efficiency of moisture use (M_e) was calculated as a function of soil moisture plus growing season precipitation used, fallow-seeded wheat used 6.7–7.0 kg·ha^{-1}·mm^{-1} and stubble-seeded wheat about 5.0 kg·ha^{-1}·mm^{-1}. In terms of precipitation received from harvest to harvest, M_e was highest in the continuous wheat rotations and lowest in the fallow rotation.

Table 23 Effect of crop rotation and method of moisture conservation on moisture conserved and moisture use efficiency on a clay loam soil in the Dark Brown soil zone at Scott, Sask.

Rotation	Snow-trapping method	Mean moisture conserved† (mm)	Mean moisture available for crop†† (mm)	Mean moisture available per yr by rotation (mm)	Mean yield wheat ($kg \cdot ha^{-1}$)		M_e ($kg \cdot ha^{-1} \cdot mm^{-1}$)	Annual M_e‡ (% of F–W)
					Fallow	Stubble		
F–W	Bare fallow	5.2c‡‡	298a‡‡	149	2036a		6.83	100
F–W	Snow-ridged fallow	20.3c	312a	156	2079a		6.66	102
F–W	7.5-m adjacent sunflower trap strip	8.7c	304a	152	2079a		6.84	102
F–W	15-m adjacent sunflower trap strip	20.3c	313a	157	2184a		6.98	107
F–W–W	Bare fallow	14.4c	307a	191	1868b		6.08	106
F–W–W	Standing stubble	69.1a	266b			1364a	5.13	
F–W	Wide-row crop (corn) fallow	42.7b	269b	135	1807b		6.71	89
Contin. W	Snow-ridged standing stubble	49.0ab	260b	260		1310a	5.04	129
Contin. W	Standing stubble	58.4ab	269b	269		1330a	4.94	131

Source: Kirkland and Keys 1981.

† Moisture conserved 1 September–1 May.

†† Moisture available, determined by moisture to 91 cm in spring minus moisture to 91 cm in fall, plus accumulated rainfall between sampling dates (1 May–1 Sept.).

‡ Calculated by (kilograms wheat per cultivated hectare)/(millimetres precipitation received).

‡‡a–c Values followed by the same letter are not significantly different at the 5% probability level as determined by Duncan's new multiple-range test.

A second continuing rotation study, being carried out on the same Elstow clay loam at Scott, was summarized for 1966–1971 and 1972–1979 by Brandt and Keys (1982). Ten crop rotations (six conventional grain and grain–forage rotations plus four moisture-enhanced F–W rotations) were examined, eight of which are shown in Table 24. Moisture enhancement in fallow involved planting in late May to early June crops such as corn, oats, sunflowers, or wheat, in rows 3.6 m apart and at right angles to the prevailing winds so as to reduce evaporation. Wheat grown on summer fallow during 1966–1971 used the most soil moisture, ranging from 105 to 125 mm (Table 24). Where wheat was grown following the moisture-enhanced, summer-fallow treatments, moisture use was similar to that for wheat grown following conventional summer fallow. Wheat grown on stubble used more moisture than oats grown for hay on stubble. Moisture used by continuous wheat was less than that for stubble-grown wheat in a fallow rotation, likely because continuous wheat produced less stubble and straw and thus trapped less snow.

During 1972–1979, canola (*B. napus* L.) grown on summer fallow used less moisture than wheat grown on the moisture-enhanced, summer-fallow treatments. Canola on fallow used slightly more moisture than wheat grown on stubble. Because canola matures in fewer days than wheat, it is generally not as heavy a user of moisture as wheat.

Residual soil moisture after wheat was consistently less than after oat hay or canola. Soil moisture after harvest of stubble-seeded wheat was similar irrespective of rotation length. Greater soil moisture after canola should increase available moisture for a succeeding crop.

In fall, soil moisture following the hay crops in the 4- to 6-year rotations was equal to, or greater than, that following wheat. In the hay rotations some accumulation of soil moisture may have occurred following breaking in July. In the 4-year rotation, soil moisture increased to 59–70 mm in the fall, whereas in the 6-year rotation soil moisture was 32–40 mm after breaking.

Storage of soil moisture during the fallow period was generally similar during 1966–1971 and 1972–1979 for all rotations; combined data for these two periods are presented in Fig. 8. Most of the soil moisture stored in fallow resulted from snowmelt during the first winter and did not vary greatly between rotations; however, storage during the summer was quite variable. Storage of summer moisture was 20–30 mm, 20 mm, and 5 mm in conventional grain, 4-year grain forage, and 6-year grain forage treatments, respectively. In the 6-year system, alfalfa was much better established than in the 4-year system and was therefore more difficult to eradicate by summer-fallowing, thereby reducing storage of soil moisture. When wheat or oats were grown in widely spaced rows to enhance the storage of moisture in summer fallow, soil moisture was lost during the summer; however, when corn or sunflowers were used, moisture storage was near normal.

Table 24 Soil moisture availability and use, Scott, Sask.

Rotation sequence	Soil moisture (mm in top 90 cm of soil)					
	Available in spring†		Used		Left	
	1966–1971	1972–1979	1966–1971	1972–1979	1966–1971	1972–1979
Continuous wheat	95	101	58	62	37	39
Summer fallow						
Wheat‡	163	162	124	101	39	61
Summer fallow						
Wheat‡	154	150	111	90	43	60
Wheat	108	113	70	74	38	39
Summer fallow						
Wheat‡	164	161	125	94	39	67
Oats	100	118	50	65	50	53
Summer fallow						
Wheat‡	162	142	121	69	41	73
Oats (alfalfa)	109	117	60	76	49	41
Alfalfa hay	104	119	34	60	70	59

(continued)

Table 24 Soil moisture availability and use, Scott, Sask. *(concluded)*

Rotation sequence	Soil moisture (mm in top 90 cm of soil)					
	Available in spring†		Used		Left	
	1966–1971	1972–1979	1966–1971	1972–1979	1966–1971	1972–1979
Summer fallow						
Wheat	142	112	105	62	37	50
Wheat	112	115	74	85	38	30
Oats (alfalfa)	111	93	64	67	47	26
Alfalfa hay	102	94	49	63	53	31
Alfalfa hay	93	93	61	53	32	40
Fallow with wheat in wide rows						
Wheat	169	139	132	101	37	38
Fallow with corn in wide rows						
Wheat	157	156	123	126	34	30

Source: Brandt and Keys 1982.
† Moisture in excess of – 1.5 MPa moisture content.
‡ This crop was spring wheat in 1966–1971 but was changed to canola in 1972–1979.

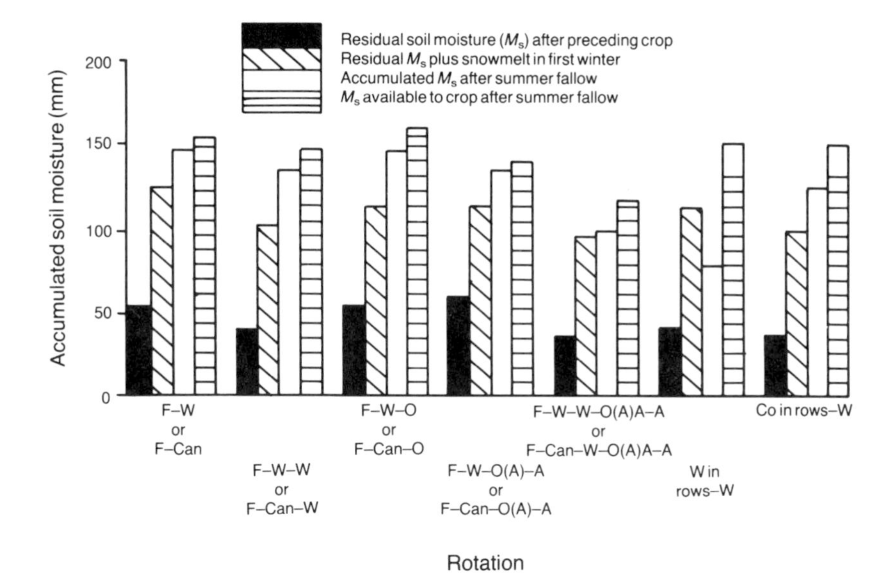

Fig. 8 Accumulation of available soil moisture (0–90 cm) in summer fallow at Scott, Sask., (adapted from Brandt and Keys 1982).

Storage of moisture from snowmelt during the second winter was highest when wheat or oats were grown in rows to enhance snow trapping. The lowest total fallow moisture storage was recorded on the 6-year rotations. The 6-year rotation also had the least quantity of available moisture for crops grown on summer fallow.

In a third study at Scott, data on yields and moisture use ($M_{sp} - M_{hv} + M_{gs}$, May 1–August 31) for two long-term rotations were analyzed by regression equations following the procedure of Campbell et al. (1988*a*). The pooled data for rotations F–W–W (both crops) and continuous wheat, both fertilized, were analyzed (S.A. Brandt, personal communication). The linear model obtained was significant ($P < 0.01$) and approximated that obtained at Swift Current:

$$y = 9.21\,(M_u - 66.23)\ (r = 0.64^{**}) \qquad [3]$$

Thus, it appears that this general regression equation can be used to estimate yields of fallow- and stubble-seeded wheat in both the Brown and Dark Brown soil zones.

Several crop rotation studies have been carried out on the Dark Brown soil at Lethbridge, Alta.; however, although soil moisture was measured, only in a few cases were these data analyzed.

Moisture data from a long-term rotation study that included F–W, F–W–W, and continuous wheat and on which various N and P treatments were superimposed starting in 1972 were subjected to regression analysis (Janzen 1986). The relationship between yields of wheat (Y_g) for the fully fertilized treatments for 1972–1984 was: $Y_g = 11.9\ (x - 50.91)$, where x was defined as available water (i.e., $M_{gs} + M_{sp}$). This equation is similar to that obtained at Swift Current and Scott, except that the coefficient of x is higher at Lethbridge.

In a more recent, short-term, crop rotation study in which grain sorghum was evaluated in six rotations from 1978–1984, sorghum exhibited no greater drought tolerance than spring wheat (Janzen et al. 1987).

Black soil zone

Few data on soil moisture have been collected, from crop rotation studies on Black soils, because post-harvest precipitation is generally sufficient to replenish soil moisture to levels equivalent to that stored in fallow soils.

In a study in Manitoba, six 3-year crop rotations were compared for 6 years on a clay loam and on a sandy loam (Spratt et al. 1975). Each rotation had two consecutive years of spring wheat following either summer fallow or one of five fallow substitute crops (i.e., biennial yellow sweetclover, potatoes, corn, flax, or oat hay). In both soils more stored moisture (measured in October) was present after summer fallow than after flax, but differences between summer fallow and the other fallow-substitute crops were small and inconsistent. In 5 of the 6 years, post-harvest precipitation was sufficient to replenish the soil moisture under fallow substitutes to the same level as that under the fallow.

A study, carried out in the Peace River region of Alberta (Hoyt and Leitch 1983), examined the effect on soil moisture reserves of legumes in cereal rotations (Fig. 9). In Black Solod and Black Solonetz soils, red clover and alfalfa depleted the moisture reserves in the subsoil; this effect persisted at depths of 60–135 cm for two succeeding crop years. However, this moisture deficit did not affect the yield of subsequent barley crops.

One of the more comprehensive studies carried out in the Black soil zone was that by Carder and Hennig (1966) on a Black Solod silt loam near Beaverlodge. They measured soil moisture daily for 5 years under F–W and red fescue ley. The results showed that only 6% of the 665 mm of precipitation received during the 20-month fallow period was stored in the soil. This amount is much less than that reported for the Brown and Dark Brown soils (18–25% of about 500 mm received). By periods, 13% of 94 mm of precipitation received between harvest and freeze-up was conserved, as was 5% of 132 mm received between freeze-up and seeding the following spring and 8% of 307 mm received between seeding and the second freeze-up; none of 132 mm received in the second winter was conserved. These low conservation efficiencies reflect the impermeable nature of the subsoils of the upper Peace River region and the consequent extensive freezing of moist soil at shallower depths in winter, which restricted infiltration of snowmelt (Carder and Hennig 1966). Not only bare fallow, but cropped systems also failed to store much of the winter precipitation; wheat stubble stored only 5% and fescue ley 10%.

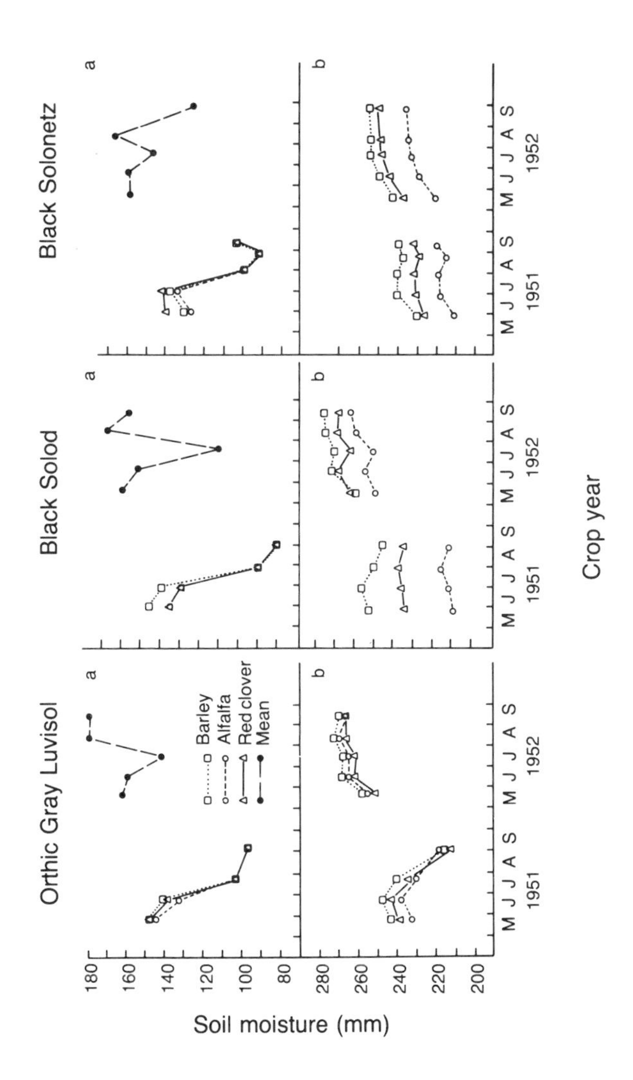

Fig. 9 The effect of previous crop in rotation on soil moisture levels during the succeeding two barley crop years (adapted from Hoyt and Leitch 1983): (*a*) at depth of 15–60 cm; (*b*) at depth of 60–135 cm.

The average amount of moisture lost from fallow by evaporation between May and freeze-up was 285 mm, i.e., only 50 mm less than that used by spring wheat and 43 mm less than that used by fescue during the same period (Figs. 10 and 11). Thus, heavy losses of moisture can occur from well-worked fallow under these moist conditions.

On average, spring wheat used 264 mm of moisture to produce 2955 $kg \cdot ha^{-1}$ of grain (i.e., M_e = 11.2 $kg \cdot ha^{-1} \cdot mm^{-1}$). Thus, M_e was much greater than for the drier soil zones. Wheat, with its more extensive root system, drew moisture from at least the 90 cm depth, whereas fescue did not draw much moisture from beyond 45 cm. However, established fescue drew moisture more rapidly and thoroughly than wheat, especially in early spring before wheat is well established. Under moist conditions, wheat used water rapidly between four-leaf (jointing) and grain-ripe stages, whereas fescue used water at a steady rate from early spring until growth stopped in fall. Harvesting fescue for hay did not greatly influence its moisture consumption.

In the Melfort area, Bowren and Townley-Smith (1986) confirmed the low efficiency of moisture conservation noted elsewhere for Black soils. They reported values ranging from 3 to 17% of the precipitation received during the fallow period was conserved in a F–W–W rotation. Moisture use efficiency for one soil was 7.8 and 7.6 $kg \cdot ha^{-1} \cdot mm^{-1}$ for fallow- and stubble-seeded wheat, respectively.

Dark Gray and Gray Luvisolic soil zones

The depletion of subsoil moisture caused by legumes was less in Gray Luvisolic soils than in the Black Solods, but this difference in depleted moisture disappeared by mid season of the first succeeding crop year (Fig. 9) (Hoyt and Leitch 1983).

An experiment, set up to determine the best time to break the forage crop, showed that the 4-year average available moisture in the top 120 cm of soil at spring seeding was 139, 134, 129, and 108 mm for July, August, September, and May breaking of fescue, respectively (Hennig and Rice 1977). However, date of breaking did not affect cereal yields showing that moisture limitation is rarely a problem in this region.

NITROGEN AND PHOSPHORUS DYNAMICS

Few of the crop rotation studies done by Agriculture Canada on the prairies have provided detailed analysis of N and P dynamics in the soil and plant.

Brown soil zone

The rate of net N mineralization in soil was quantified as a function of precipitation in the crop rotation study at Swift Current

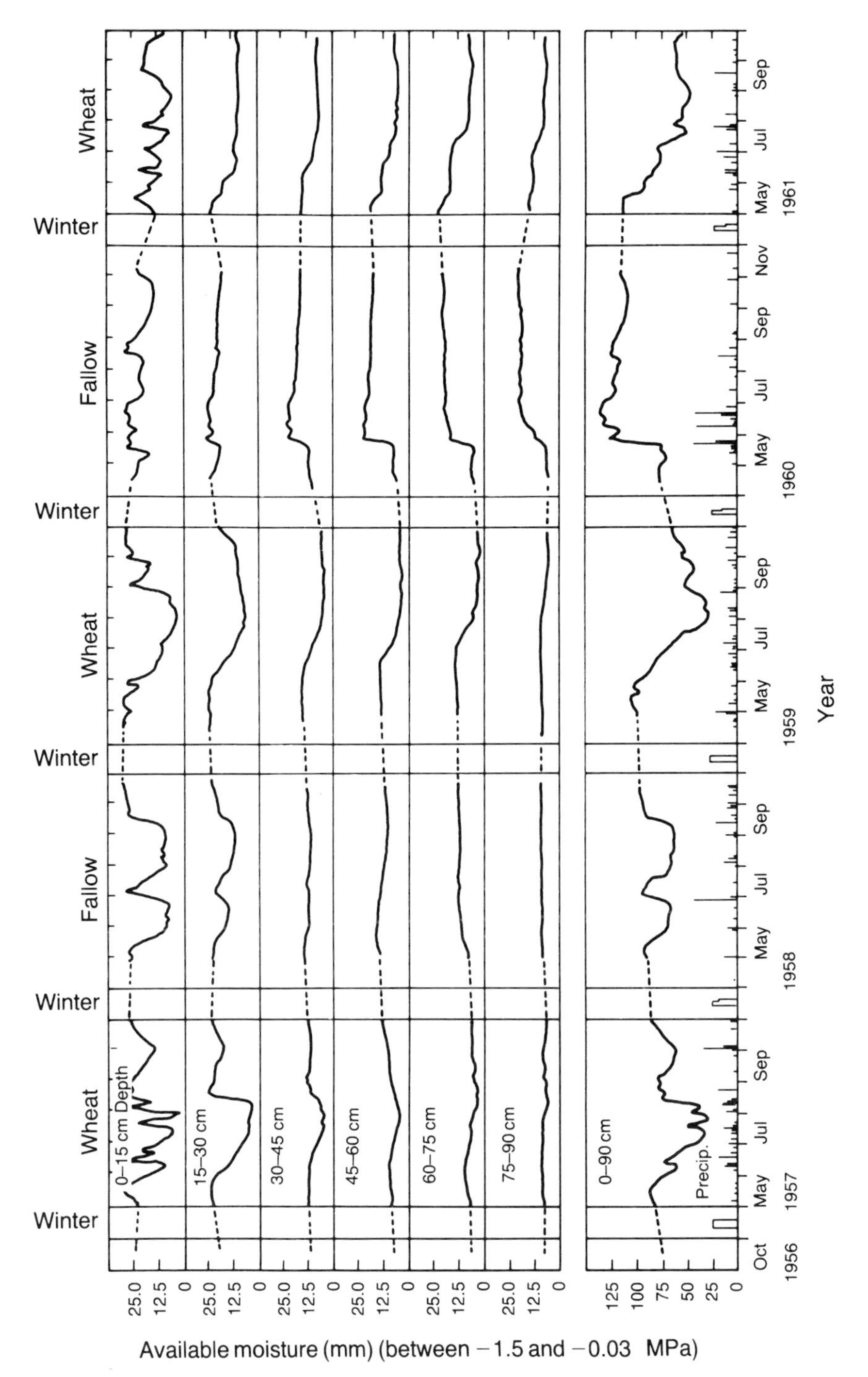

Fig. 10 Moisture regime under a wheat–fallow rotation on a Black soil in the Peace River region of Alberta (redrawn from Carder and Hennig 1966).

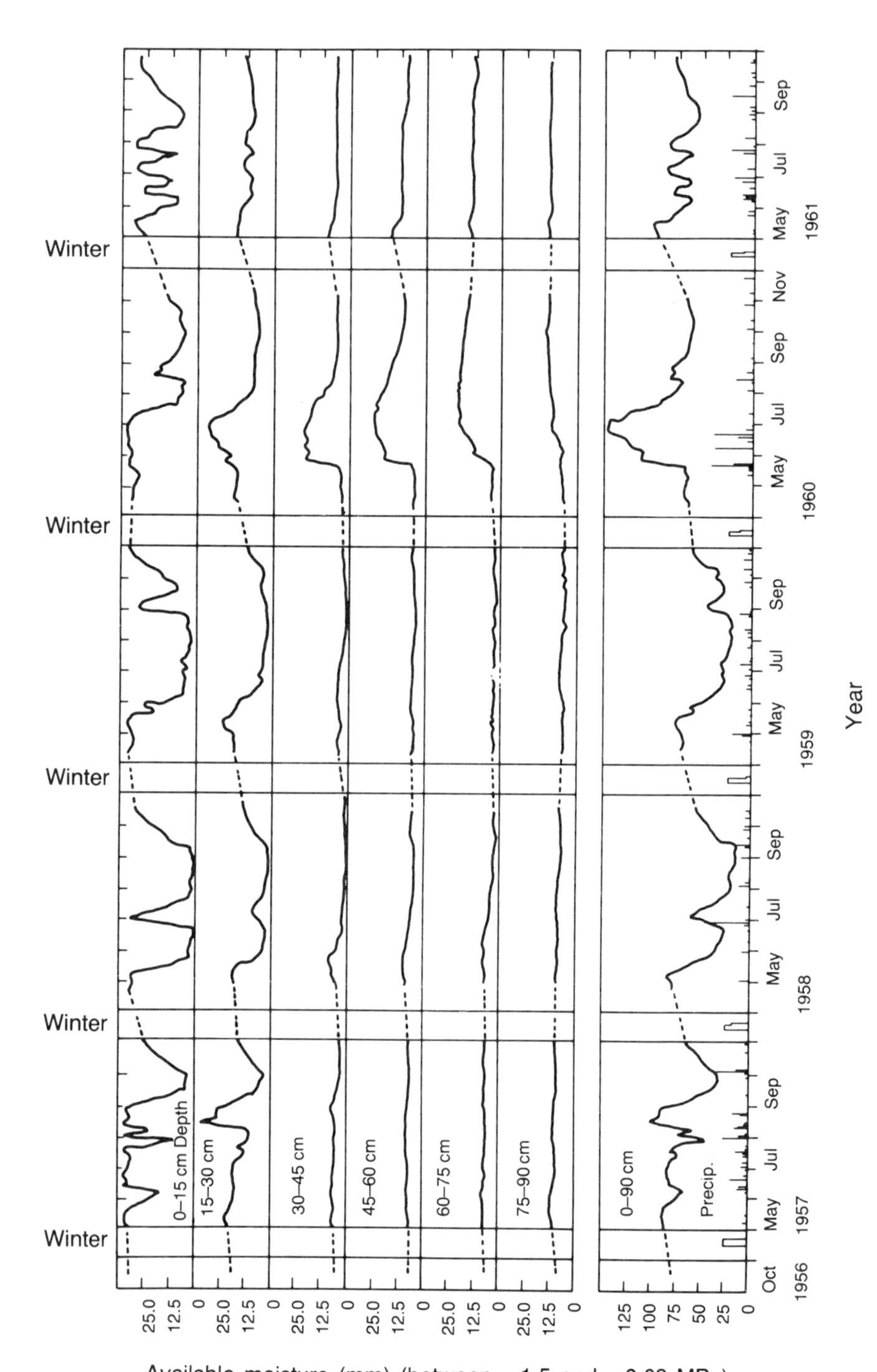

Fig. 11 Moisture regime under a red fescue ley on a Black soil in the Peace River region of Alberta (redrawn from Carder and Hennig 1966).

(Campbell et al. 1983*a*, Campbell et al. 1988*b*). Based on data for the first 18 years, the relationship between net N mineralized (N_{min} in $kg \cdot ha^{-1}$) in the 0–60 cm depth of fallow soil and precipitation (mm) between early May and mid October was represented by the equation: $N_{min} = 23 + 0.29$ precip, ($r = 0.75^*$) (Campbell et al. 1988*b*). The equation suggests that, if we exclude leaching and other major loss mechanisms, net N_{min} will increase in this Swinton loam, when summer-fallowed, by about 0.0018 $kg \cdot ha^{-1} \cdot mm^{-1}$ precipitation received from spring to fall. Based on N balance approach, net N mineralized under the crop was about 20–30 $kg \cdot ha^{-1}$ on average, but values were as high as 95 $kg \cdot ha^{-1}$.

The distribution of nitrate-nitrogen (NO_3-N) in the soil and N uptake by the crop during the first 12 years of the Swift Current rotation were studied by monitoring of soil and plant N (Campbell et al. 1983*a*). Nitrogen was found in amounts of between 38–54 $kg \cdot ha^{-1}$ in the grain of spring wheat each year. The well-known, gradual buildup of NO_3-N in fallow soil and draw-down from N uptake by crops was demonstrated. Also NO_3-N was being leached beyond the rooting zone of the cereal crops in years of above-average precipitation and even in some relatively dry years when heavy spring rains occurred over short periods (Campbell et al. 1983*a*). This finding was later confirmed during the 1982 growing season when above-average rainfall leached N at 123 $kg \cdot ha^{-1}$ beyond the root zone (Campbell et al. 1984*a*). Leaching was observed under continuous wheat, although the losses were much less than those in rotations with frequent fallow (Fig. 12). Replacing spring wheat with fall rye minimized leaching losses because of its early growth initiation and its high consumption of water in the spring. Proper fertilization of the crop reduced subsoil nitrate by about 20%, probably because of enhanced water and N consumption by the fertilized crop. In wet years good plant growth resulted in large N uptake, thus reducing nitrate leached, but in dry years the amount of nitrate in the 60–120 cm depth remained high throughout the growing season (Campbell et al. 1983*a*).

Analysis of the P data from the Swift Current study after 12 years showed that the level of bicarbonate-soluble P (labile P_i) was not significantly changed in a treatment receiving no P fertilizer (Campbell et al. 1984*b*). However, frequent application of P fertilizer at generally recommended rates increased the bicarbonate-soluble P_i in soils 0–15 cm deep, but not those at lower depths. Surprisingly, changes in the P_i fraction of fallowed soil appeared to be correlated with changes in cropped treatments ($r = 0.62^{**}$ to 0.85^{**}), which is in contrast to the behavior of NO_3-N in soil. This finding was confirmed by O'Halloran (1986) in a more detailed investigation of these rotations.

---

* Significant at $P < 0.05$.

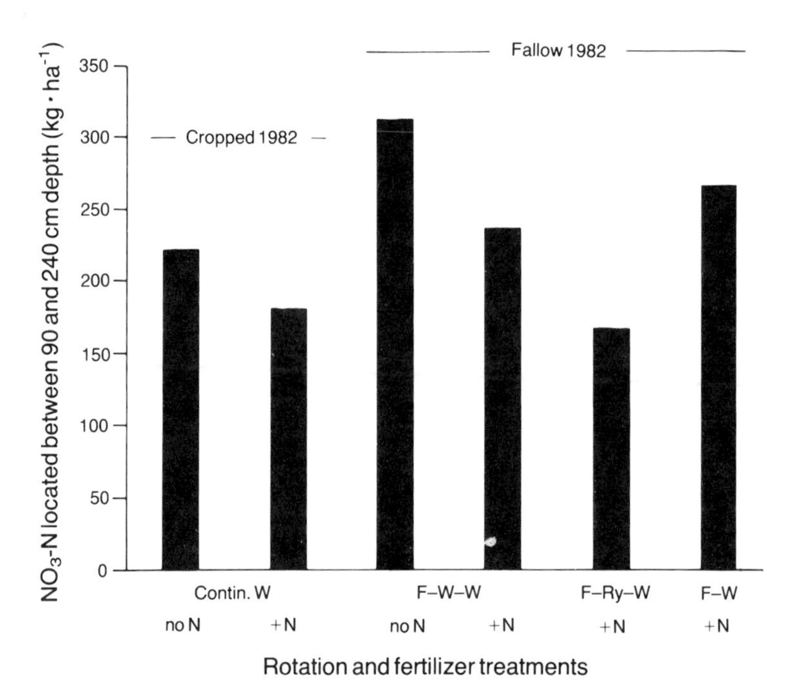

Fig. 12 Subsoil NO_3-N in Brown Chernozem at Swift Current, Sask., as influenced by crop rotation and fertilizer application (redrawn from Campbell et al. 1984*a*).

Campbell et al. (1984*b*) and O'Halloran (1986) observed that labile P_i changes during the period of summer fallow were related significantly to soil moisture and temperature. O'Halloran (1986) also found that annual additions of P fertilizers increased the labile and moderately labile P_i fractions of the soil but did not affect the labile and moderately labile organic P fractions (P_0 forms). The labile and moderately labile P_0 forms were increased in the presence of a crop, but only at lower ($<32\%$) sand contents (i.e., higher silt plus clay contents). He further observed that annual additions of P fertilizers and continuous wheat cropping increased the levels of more stable P_0 forms in the soil (at sand contents $<28\%$) as compared to a F–W–W rotation with no additional P fertilizer.

Campbell et al. (1984*b*) reported that bicarbonate-soluble P_i increased between late fall and spring thaw. Phosphorus uptake by plants increased with time up to shot-blade stage; the rate slowed thereafter, but uptake continued to maturity. The amount of P uptake was directly related to the yield of dry matter and positively influenced by N fertilizer.

Dark Brown soil zone

No information on net N or P mineralization, or on N and P uptake, was reported for the rotation studies carried out at Scott and Lethbridge. Considerable work of this nature was initiated recently at Lethbridge (Stewart et al. 1989).

Black and Gray soil zones

Measurements of the effects both of legumes in rotation and of breaking sod on N and P were determined at Brandon, Indian Head, and Beaverlodge.

In a 6-year study carried out on an Assiniboine clay loam and a Miniota sandy loam (Black soils) in Manitoba, Spratt et al. (1975) examined the effect of the following five summer-fallow substitute crops on the yield of spring wheat in a FS–W–W rotation. The FS crops were yellow sweetclover, potatoes, corn, oat hay, and flax. They found that the NO_3-N, measured at seeding of the first wheat crop, was higher in fallow and under sweetclover than under the other treatments, but this difference did not carry over to seeding of the second wheat crop. Although differences in spring NO_3-N levels were not reflected in yields they were related directly to grain protein.

Ferguson and Gorby (1971) conducted a 12-year study on a Marringhurst sandy loam, a Black soil at Brandon, Man., to determine the effect that including alfalfa, bromegrass, or bromegrass–alfalfa, for different periods, would have on the subsequent yields of wheat. Three cereal systems (viz., grain–grain–fallow, grain–fallow, and continuous fallow) were used as controls (Table 25). The amount

Table 25 Total nitrogen in the above-ground portion of wheat grown on pre-treated land (measured at the soft-dough stage)

		N ($kg \cdot ha^{-1}$)			
Pre-treatment	Rotation	1966	1967	1968	Total
Grain rotation	F–W–W	71	31	49	151
	F–W	69	28	41	138
	Contin. F	64	24	39	127
4-yr seed-down	A	85	43	59	187
	Br–A	89	40	59	188
	Br	75	31	48	154
6-yr seed-down	A	95	49	69	213
	Br–A	90	43	67	200
	Br	81	44	57	182
8-yr seed-down	A	98	54	80	232
	Br–A	90	50	72	212
	Br	93	46	67	206
Least significant difference ($P < 0.05$)		16	8	13	

Source: Ferguson and Gorby 1971.

of available N in the 0–60 cm depth after fallowing the soil, as reflected by yield and N uptake responses, was highest when alfalfa was included in the system (Table 25), next highest when bromegrass was included, and lowest for the cereal-only system (Fig. 13) (Ferguson and Gorby 1971).

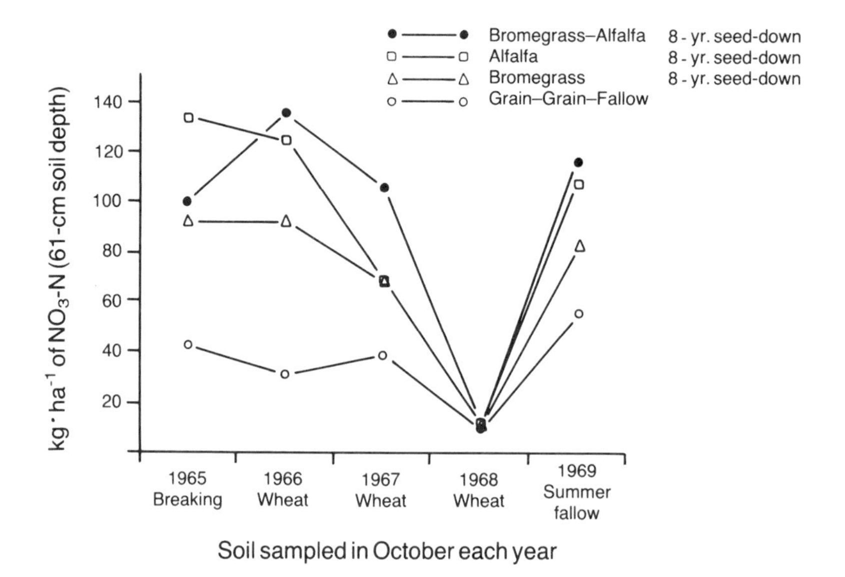

Fig. 13 The influence of previous crops on the NO_3-N content of the top 61 cm of soil after wheat and summer fallow (redrawn from Ferguson and Gorby 1971).

At Beaverlodge, Hoyt and Leitch (1983) assessed the availability of soil N left by five legumes for the succeeding barley crop grown on three Gray Luvisols and two Black Solodic soils. Although one of the Black soils (Rycroft) showed a decrease in available residual N, the other soils showed a substantial increase (Fig. 14). It was suggested that this residual N was derived from the decomposition of legume roots. Benefits of the residual N to the crops was apparent in the initial years after breaking and then the benefits diminished, but were still apparent even after the seventh successive crop (Fig. 15).

The amount of available N produced for the cereal crop after breaking a fescue ley was found to be directly related to the interval between breaking of sod and seeding of the subsequent crop, which was reflected in N uptake by the crop (Hennig and Rice 1977). However, if the soil was fallowed for a further 12 months (i.e., for the entire 21-month fallow period) there was no treatment effect of NO_3-N because more N mineralization occurred in the treatments that were broken later.

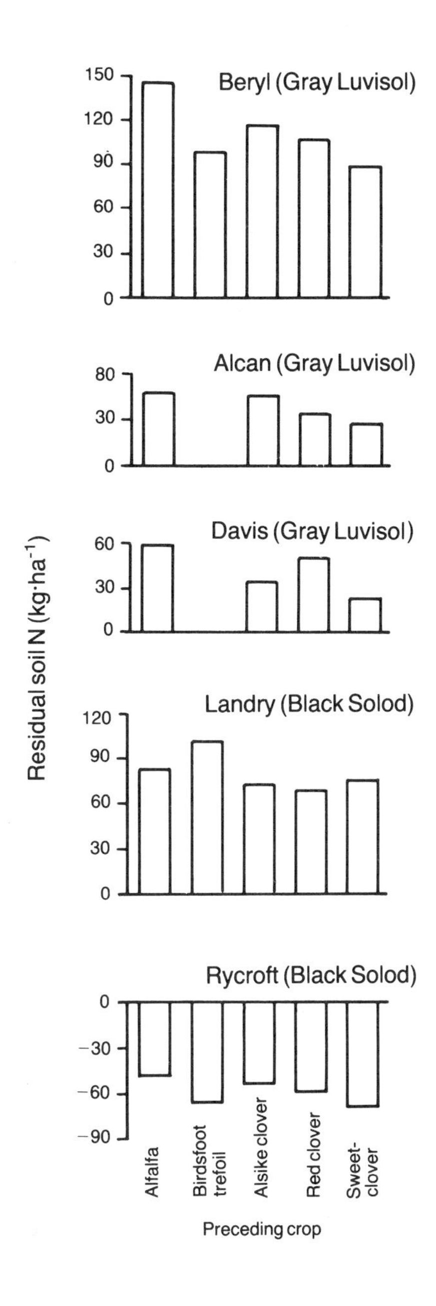

Fig. 14 Availability of residual soil N to the succeeding barley crops in five soils following various legumes (redrawn from Hoyt and Leitch 1983).

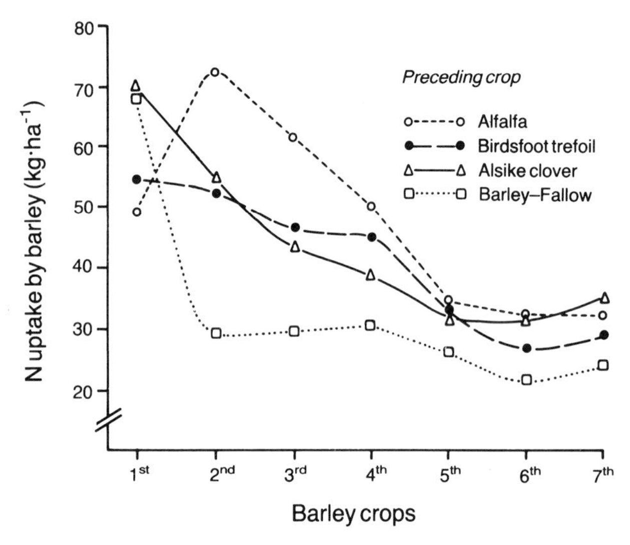

Fig. 15 N uptake in seven crops of barley following legumes or a barley–fallow control on Beryl soil (redrawn from Hoyt and Leitch 1983). (Results for red clover were similar to those for alsike clover.)

SOIL QUALITY

The soils of western Canada have undergone serious degradation since arable agriculture began in the early part of the 20th century (McGill et al. 1981). Of particular concern has been depletion of quantity and quality of the soil's organic matter. Concern about soil degradation has led to investigation to identify rotations that preserve soil quality over the long term. Crop residues are the primary substrate for replenishment of organic matter; thus changes in crops and their sequence in rotation can influence soil quality significantly.

Swift Current and Lethbridge carried out detailed research on the effect of crop rotations on soil quality. Variables that were examined include the following: fallowing frequency, inclusion of alternative crops (oilseeds, legumes, and forages), the addition of fertilizer amendments, and the influence of time of cultivation.

Brown soil zone

After 15 years of cropping to various rotations at Swift Current, soil C and N concentrations were about 17% higher in continuous wheat than in F–W or F–W–W treatments (Table 26) (Biederbeck et al. 1984). The effect on organic matter quality was even more pronounced with mineralizable N levels in the continuous wheat soil being more than 40% higher than in the F–W soil.

At Swift Current, the number of bacterial heterotrophs in the soil and microbial biomass C and N increased as the frequency of fallow decreased (Table 27) (Biederbeck et al. 1984). Crop rotation affected

Table 26 The influence of fallowing frequency on organic C content, total N content, C:N ratio, potentially mineralizable N content (N_o), and potentially mineralizable N as a proportion of total N

Location and duration	Depth (cm)	Rotation sequence	Organic matter characteristics					Fertilizer	Source
			Organic C (%)	Total N (%)	C:N ratio	N_o (mg·kg−1)	N_o: total N ratio (%)		
Brown soil zone									
Swift Current 1967–1982	0–7.5	Contin. **W**	2.15	0.226	9.5	230	10.2	N,P	*a*
		F–W–W	1.70	0.189	9.0	197	10.4	N,P	
		F–W	1.94	0.202	9.6	159	7.9	N,P	
Dark Brown soil zone									
Lethbridge 1912–1975	0–13	Contin. **W**	1.91	0.17	11.2	–	–	none	*b*
		F–W–W	1.55	0.15	10.6	–	–	none	
		F–W	1.36	0.12	10.9	–	–	none	
Lethbridge 1951–1984	0–15	Contin. **W**	1.78	0.180	9.9	133	7.4	none	*c*
		F–W–W	1.58	0.160	9.8	105	6.5	none	
		F–W	1.50	0.156	9.6	68	4.3	none	

(continued)

Table 26 The influence of fallowing frequency on organic C content, total N content, C:N ratio, potentially mineralizable N content (N_o), and potentially mineralizable N as a proportion of total N (*concluded*)

Location and duration	Depth (cm)	Rotation sequence	Organic matter characteristics: Organic C (%)	Total N (%)	C:N ratio	N_o (mg·kg−1)	N_o: total N ratio (%)	Fertilizer	Source
Black soil zone									
Indian Head 1958–1984	0–7.5	Contin. **W**	2.43	0.198	12.3	–	–	none	*d*
		F–W–W	2.25	0.186	12.1	–	–	none	
		F–W	2.21	0.179	12.3	–	–	none	
Melfort 1957–1984	0–7.5	Contin. **W**	5.92	0.532	11.1	–	–	none	*d*
		F–W–W	5.30	0.501	10.6	–	–	none	
Solonetzic soil									
Vegreville 1958–1980	(Ap)	Contin. **W**	4.24	0.34	12.5	55	2.2	N,P	*e*
		F–W	3.79	0.25	15.7	95	2.8	N,P	

Sources: (*a*) Biederbeck et al. 1984, (*b*) Dormaar and Pittman 1980, (*c*) Janzen 1987*b*, (*d*) Campbell (unpublished), and (*e*) Carter 1984.

the biomass composition as well. For example, the ratio of bacteria to actinomycetes in the F–W rotation was severalfold smaller than that in continuous wheat treatments. A further indication of differences in microbial populations was a widening of microbial C:N ratios from the F–W to continuous wheat treatments, indicative of a progressively greater predominance of fungal over bacterial populations. Soil respiration, which is indicative of microbial activity, was inversely related to fallowing frequency (Table 27).

The activity of several enzymes, including dehydrogenase and urease, increased significantly with progressive decreases in fallowing frequency (Biederbeck et al. 1986, Biederbeck and Campbell 1987). Thus, urease activity increased from 44 activity units in a fertilized F–W rotation (mean of two phases) to 75 activity units in a fertilized continuous wheat treatment. The activity of other enzymes, such as arylsulfatase and two phosphatases, were not appreciably affected by fallow frequency. The total concentration of amino acids in the soil increased progressively from the F–W to the continuous wheat treatments, but the relative proportions of various amino acids was not affected.

Table 27 Effect of fallowing frequency and fertilizer application on biological properties in the 0–7.5 cm depth of a fertilized Swinton loam at Swift Current, Sask., after 16 years of cropping

Property	Rotation and fertilizer treatments			
	F–W (N + P)	F–W–W (N + P)	Contin. W (N + P)	Contin. W (P)
Microbial counts (org·g^{-1} soil)				
Bacteria ($\times 10^6$)	82	135	156	201
Actinomycetes ($\times 10^6$)	34	37	24	26
Bacteria:actinomycete ratio	2.4	3.7	6.5	7.8
Yeasts ($\times 10^4$)	11	12	24	15
Microbial biomass				
Biomass C (mg·kg^{-1})	216	271	260	428
Biomass C (% of soil C)	1.1	1.6	1.2	2.4
Biomass N (mg·kg^{-1})	62	78	65	86
Biomass N (% of soil N)	3.1	4.1	2.9	4.4
Biomass C:N ratio	3.4	3.5	4.1	5.1
Respiration (mg·kg^{-1} of CO_2-C in 10 days at 20°C)	92	99	138	111

Source: Adapted from Biederbeck et al. 1984.

A soil characteristic that has important implications for long-term productivity is erodibility. Any inherent soil property that reduces susceptibility to erosion is likely to enhance long-term soil quality. Biederbeck et al. (1984) demonstrated that potential soil erodibility (fraction of soil with aggregate diameter <0.84 mm as determined by dry-sieving) was markedly higher in F–W rotations than in a continuous wheat system (Fig. 16). Trash cover, which is another index of a soil's susceptibility to erosion, was twice as high in continuous wheat as in a F–W rotation. Thus, frequent inclusion of fallow in the rotation reduces trash cover and encourages erosion.

Several noncereals have only become important economic crops in recent decades, thus their long-term effects on soil productivity have not been fully established. Biederbeck et al. (1984) studied two rotations involving flax (F–Flx–W and Flx–W–W) in a comparison with other rotations involving spring wheat and fallow. Replacement of wheat with flax appeared to have little influence on contents of organic matter, and the overriding influence remained the effect of fallowing frequency. Analysis of trash conservation and aggregate size distribution suggested that the erodibility of soil in the Flx–W–W rotation was slightly greater than that of continuous wheat (Fig. 16), perhaps reflecting the lower amount of residue produced by flax relative to wheat.

Nitrogen fertilization for 16 years significantly enhanced content and quality of organic matter in the soil (Biederbeck et al. 1984). Carbon and nitrogen concentrations were 21 and 15% higher, respectively, in a continuous wheat treatment receiving both N and P than in a continuous wheat treatment receiving only P; the potentially mineralizable N concentration was 24% higher in the treatment receiving both N and P (Table 28). This enhancement of soil quality can probably be attributed to increased amounts and higher nutrient contents of crop residues applied to the soil.

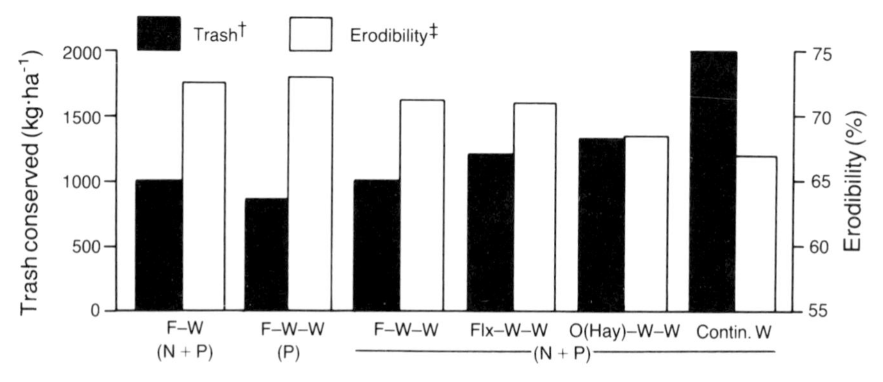

†Based on average of five 1-m^2 samples per plot.
‡Based on % of soil diameter <0.84 mm; samples taken from top 5 cm of soil.

Fig. 16 The influence of crop rotation and fertilizer application on erodibility (percentage of soil with aggregate diameter <0.84 mm) and trash conservation in a Brown Chernozemic soil at Swift Current, Sask., (redrawn from Biederbeck et al. 1984).

Table 28 The influence of fertilizer application on organic C and total N contents, C:N ratio, potentially mineralizable N content (N_0), and potentially mineralizable N as a proportion of total N

			Organic matter characteristics						
Location and duration	Depth (cm)	Rotation sequence	Fertilizer treatment	Organic C (%)	Total N (%)	C:N ratio	N_0 (mg·kg−1)	N_0: total N ratio (%)	Source
Brown soil zone									
Swift Current 1967–1982	0–7.5	Contin. **W**	P	1.78	0.197	9.0	185	9.4	*a*
			N + P	2.15	0.226	9.5	230	10.2	
Dark Brown soil zone									
Lethbridge 1912–1975	0–13	**F**–W–W	none	1.43	0.141	10.3	151	10.7	*b*
			P	1.46	0.138	10.6	140	10.1	
			N	1.67	0.155	10.7	207	13.4	
			N + P	1.64	0.141	10.6	213	15.1	
		Contin. **W**	none	1.62	0.149	10.9	192	12.9	
			P	1.61	0.141	11.5	221	15.7	
			N	1.80	0.162	11.1	255	15.7	
			N + P	1.88	0.171	11.0	250	14.6	

(continued)

Table 28 The influence of fertilizer application on organic C and total N contents, C:N ratio, potentially mineralizable N content (N_o), and potentially mineralizable N as a proportion of total N (*concluded*)

Location and duration	Depth (cm)	Rotation sequence	Organic matter characteristics: Fertilizer treatment	Organic C (%)	Total N (%)	C:N ratio	N_o (mg·kg−1)	N_o: total N ratio (%)	Source
Black soil zone									
Indian Head 1958–1984	0–7.5	**F–W**	none	2.21	0.179	12.3	–	–	*c*
			N + P	2.28	0.186	12.2	–	–	
		F–W–W	none	2.25	0.186	12.1	–	–	
			N + P	2.38	0.200	11.9	–	–	
		Contin. **W**	none	2.43	0.198	12.3	–	–	
			N + P	2.59	0.223	11.6	–	–	
Melfort 1957–1984	0–7.5	**F–W–W**	none	5.30	0.501	10.6	–	–	*c*
			N + P	5.04	0.498	10.1	–	–	
		Contin. **W**	none	5.92	0.527	11.2	–	–	
			N + P	6.02	0.532	11.3	–	–	

Sources: (*a*) Biederbeck et al. 1984, (*b*) Janzen 1987*b*, (*c*) Campbell (unpublished).

Several biological properties were also affected significantly by the application of N fertilizer to the continuous wheat treatment at Swift Current (Table 27). Bacterial numbers as well as microbial biomass N and C concentrations were appreciably higher in the treatment receiving only P than in that fertilized with both N and P. Conversely, soil respiration, a measure of microbial activity, was significantly higher in the treatment fertilized with both N and P than in that fertilized only with P. This apparent anomaly was attributed to the presence of a relatively large but inactive microbial population in the N-starved conditions associated with growing continuous wheat without N fertilizer.

Fertilization also increased the production of the crop's biomass and, consequently, the amount of residue returned to the soil (Campbell et al. 1988*b*). Also fertilization may reduce susceptibility to erosion by increasing soil aggregation (Biederbeck et al. 1984).

A potentially negative effect of long-term reliance upon inorganic fertilizers in crop rotations is the acidification associated with frequent use of ammonium-based N fertilizers. Campbell and Zentner (1984) reported an average pH decline of 0.5 units as a result of 17 years of ammonium nitrate application (supplying N at 35 $kg \cdot ha^{-1}$) to a continuous wheat crop rotation.

Biederbeck et al. (1984) observed a consistent trend for increasing the concentration of organic N in surface soil in all the various spring wheat rotation treatments, though the size of that increase was influenced by rotation and fertilizer treatment (Fig. 17). These trends for increased organic N concentration occurred despite apparent N deficits (greater removal than addition of inorganic N). The authors hypothesized that the "extra" N occurred as a result of N uptake from deep within the soil profile.

Dark Brown soil zone

At Lethbridge, an inverse relationship was found between fallowing frequency and the concentration of organic matter in the soil (Table 26). The degradative effect of fallowing was more pronounced on measurements of soil quality than on total organic matter (Fig. 18 and Table 26).

Janzen (1987*b*) observed that soil organic N concentrations in a continuous wheat treatment were 15% more than those in a F–W rotation. In the same experiment, mineralizable N concentrations in the continuous wheat rotation were about twice those in the fallow rotation. The concentration of potentially mineralizable N appeared to be related to levels of partially decomposed organic substrate ("light fraction" organic matter) in the soil.

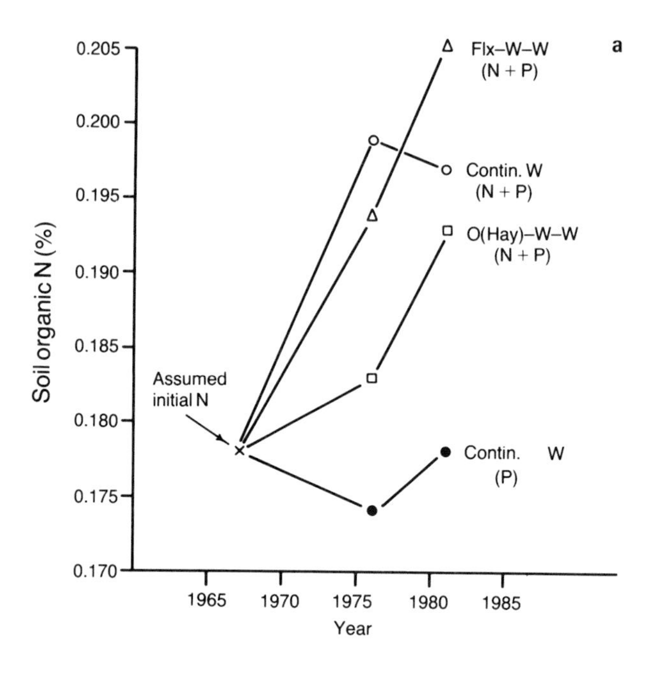

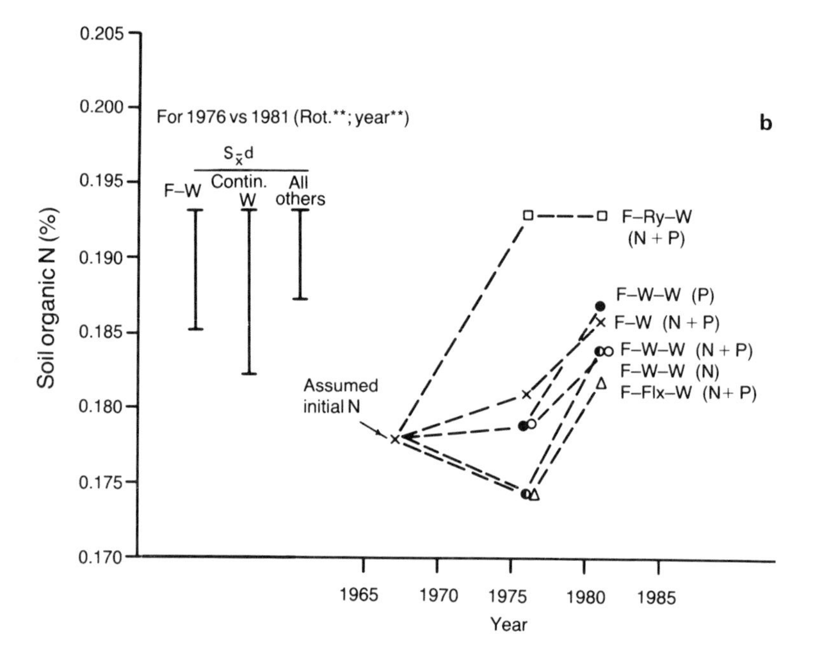

Fig. 17 Changes in soil organic N at depths from 0–15 cm during 15 years of cropping to various rotations at Swift Current, Sask., for (*a*) continuous cropping rotations and (*b*) fallow rotations (redrawn from Biederbeck et al. 1984). ($S_{\bar{x}}d$ represents the standard error of the difference between treatment means.)

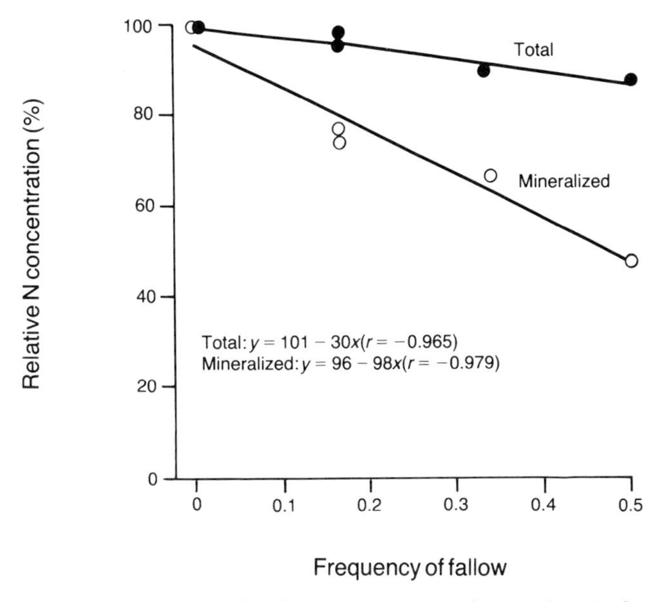

Fig. 18 Influence of fallowing frequency on relative total and mineralizable N concentrations in a Dark Brown Chernozemic soil after 31 years of cropping to various rotations at Lethbridge, Alta., (redrawn from Janzen 1987*b*).

Soil respiration decreased with progressive increase in the frequency of fallow in a rotation study sampled 33 years after initiation (Janzen 1987*b*). Respiration in the continuous wheat treatment was about twice that in the F–W rotation. Thus, potential microbial activity in soil is enhanced by reducing the frequency of fallow in the rotation.

Results similar to those obtained at Swift Current were reported by Dormaar (1983), who found an inverse relationship between the frequency of fallow and the activity of dehydrogenase, phosphatase, and urease enzymes. For example, dehydrogenase values (averaged across rotation phases) were 46, 31, and 24 nmol of formazan released hourly per gram of soil in the continuous wheat, F–W–W, and F–W rotations, respectively. A high frequency of fallow in the rotation reduced the monosaccharide content of the soil, particularly in small aggregates (Dormaar 1984).

Studies at Lethbridge (Dormaar and Pittman 1980, Dormaar 1983) revealed appreciable declines in aggregate stability with increasing frequency of fallow in spring wheat rotations; furthermore the surface soil of continuous wheat had twice as much crop residue as F–W. After 65 years of cropping, the proportion of soil aggregates stable in water, averaged across rotation phases, was 46, 36, and 35% in continuous wheat, F–W–W, and F–W rotations, respectively. Dormaar (1983) further reported a much higher content of long-chain, aliphatic carboxylic acids in continuous wheat than in F–W soil. These acids are believed to promote the aggregation of soil particles.

At Lethbridge, Janzen (1987*b*) compared the effect of including perennial forage (alfalfa and crested wheatgrass) in a 6-year cereal–forage rotation (F–W–W–H–H–H) with that of several monoculture wheat rotations. The forage rotation was more effective than F–W and F–W–W rotations in conserving soil organic C and N. After 31 years of cropping, however, soil organic matter in the forage rotation was no higher than that of the continuous wheat rotation. Similarly, concentrations of mineralizable N in the forage rotation were higher than those in the F–W and F–W–W rotations but not higher than that of the continuous wheat treatment. The absence of a significant benefit from the inclusion of forages relative to continuous wheat was attributed to the high soil-conserving value of continuous wheat, adverse effects of the fallow in the forage rotation, and limitations imposed on the perennial forage by moisture stress prevalent in the Dark Brown soil zone.

Pittman (1977) reported that the inclusion of alfalfa–crested wheatgrass in the 6-year forage–wheat system reduced the rate of soil N depletion but accelerated slightly the decline in exchangeable potassium levels.

Fertilizer affected soil quality in a Dark Brown soil at Lethbridge similarly to the Brown soil zone (Table 28). Application of N fertilizer (providing nitrogen at 45 $kg{\cdot}ha^{-1}$) for 18 consecutive years increased concentrations of soil organic C and N by about 15% over treatments receiving no N fertilizer in continuous wheat and F–W–W rotations (Janzen 1987*a*). Phosphorus application had no significant effect on organic matter, probably because of relatively high levels of indigenous soil P and, hence, little yield response to P. Fertilizer affected the labile organic matter fractions much more than total concentrations of organic matter. Potentially mineralizable N levels were increased by an average of 33% in N-fertilized soil compared with soil receiving no N fertilizer; P fertilizer had no significant effect.

Nitrogen fertilization significantly enhanced soil respiration in the continuous wheat and F–W–W systems (Janzen 1987*a*). These results, along with those at Swift Current, suggest an enhancement of soil microbial activity by N fertilization, presumably because of increased plant growth and thus greater additions of organic substrates.

Janzen (1987*a*) found a 0.3-unit decline in pH after 18 years applications of ammonium nitrate providing N at 45 $kg{\cdot}ha^{-1}$ to continuous wheat or F–W–W at Lethbridge.

The long-term study at Lethbridge affords a unique opportunity to measure temporal changes in soil quality because the experiment was established immediately after the original breaking of the soil and has been sampled periodically since then. Freyman et al. (1982) reported a rapid decline in organic C concentration in the first 12 years following initial cultivation in 1910, but a gradual increase from then on (Fig. 19). By 1980, concentrations of organic C approached original values. The resurgence of organic C levels can probably be attributed

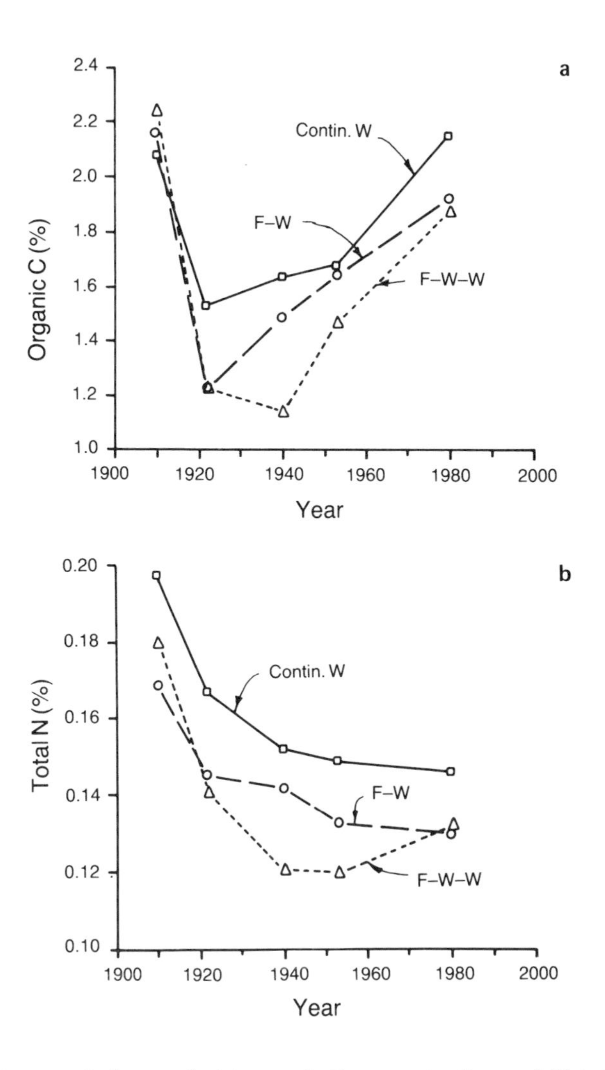

Fig. 19 Temporal changes in (*a*) organic C concentration and (*b*) total N concentration in an unfertilized Dark Brown Chernozemic soil at Lethbridge, Alta., as influenced by crop rotation (redrawn from Freyman et al. 1982).

to increased additions of crop residues resulting from improved crop yields and soil management practices. One important factor may be the introduction of combine harvesters, which, unlike early threshing methods, returned all straw to the soil. Soil N concentrations declined rapidly in the first three decades of cultivation but stabilized and remained relatively constant thereafter despite continual removal of N in the grain (Fig. 19). The maintenance of soil N contents, even in the absence of fertilizer application, suggests significant input of N from external sources, including the possible uptake of N from deep within the soil profile, accession of N in precipitation, and biological fixation of atmospheric N.

Black soil zone

Results of a long-term rotation experiment at Indian Head demonstrated the degradative effect of fallowing on soil organic matter (Table 26). After 26 years of cropping, concentrations of soil organic C and N were about 10% higher in an unfertilized continuous wheat treatment than in a comparable F–W rotation. Similarly, at Melfort concentrations of organic C and N in a Black soil receiving no fertilizer were appreciably higher in a continuous wheat treatment than in a F–W–W rotation (Table 26). The magnitude of the relative increase after 27 years of cropping was 12 and 6% for organic C and N, respectively. The relationship between fallowing frequency and concentration of organic matter was not as consistent in fertilized treatments where organic C concentrations were higher in F–W and continuous wheat treatments than in the F–W–W rotation. In comparison, at Brandon, Ferguson and Gorby (1971) found no significant difference in soil N concentrations after 12 years of cropping to various treatments ranging in fallowing frequency from every year to every 3rd year. In other studies, largest differences in concentrations of organic matter were normally observed between the continuous wheat and other wheat rotations that included fallow.

Spratt (1966) found appreciable advantage to the inclusion of forage in wheat rotations at Indian Head. This experiment compared two rotations: F–W–W and a 9-year forage-containing rotation that also received barnyard manure. After 50 years of cropping, the forage rotation improved the physical soil properties, relative to those in the F–W–W rotation, as indicated by measurements of bulk density, pore-size distribution, porosity, aggregate size distribution, and organic matter. However, because the forage rotation received manure whereas the F–W–W did not, interpretation was difficult.

In a 12-year study at Brandon, Ferguson and Gorby (1971) noted that losses of soil N were lower when the cropping system included various forage crops for a portion of the cropping period as compared to when the soil was cropped to various wheat and fallow treatments. The proportional loss of soil N over 12 years was 14% for a treatment including 8 years of various forages (alfalfa, bromegrass, or mixture of the two) compared to a loss of 20% or more for various wheat rotations including from 33 to 100% fallow.

A study initiated at Indian Head in 1958 included a 6-year rotation involving 3 years of bromegrass–alfalfa forage, 1 year of fallow, and 2 years of spring wheat. The forage rotation maintained higher levels of organic C and N in the surface soil (0–7.5 cm) than either F–W or F–W–W rotations. Concentrations of organic N and C in the forage rotation soil were higher than those in unfertilized continuous wheat but not significantly higher than those in a fertilized continuous wheat. For example, total concentrations of N in the surface soil for the F–W, F–W–W, continuous wheat, and forage rotation were 0.186, 0.200, 0.223, and 0.229%, respectively, when the first three rotations were all fertilized. The same experiment also included a GM–W–W rotation in which a crop of sweetclover was grown as green manure and plowed down during the fallow year. Concentrations of organic C and N in this latter rotation were comparable to the fertilized F–W–W rotation and lower than those observed for either fertilized continuous wheat or the 6-year forage rotation.

An experiment at Melfort included a 6-year forage–cereal rotation with 1 year of fallow, 3 years of wheat, and 2 years of bromegrass–alfalfa forage cut for hay. Concentrations of organic C and N in the surface soil of this rotation sampled at various phases were comparable to those observed in continuous wheat treatments and somewhat higher than those in F–W–W treatments. A 3-year rotation of GM–W–W, in which the green manure crop was sweetclover, exhibited concentrations of organic C and N similar to those in the fertilized and unfertilized conventional F–W–W rotations but less than those in the continuous wheat and the 6-year forage–cereal systems (data not shown).

The application of N and P fertilizer was shown to enhance concentrations of organic C and N in the surface soil of the rotation experiment at Indian Head (Table 28), although not all differences were statistically significant. The benefit of fertilizer application appeared to increase with progressive decline in fallowing frequency. Thus increases in organic C concentrations amounted to 3, 6, and 7% in the F–W, F–W–W, and continuous wheat treatments, respectively. The corresponding increases in total N concentration were 4, 8, and 13%, respectively. This interactive effect of fertilizer application and crop rotation can probably be attributed to higher yield responses in wheat grown on stubble than in wheat grown on fallow.

Application of N and P fertilizer had inconsistent and generally little effect on the content of organic matter of a Black soil at Melfort (Table 28). The absence of significant benefit of fertilizer application to the organic matter content of this soil, in contrast to results observed in other soil zones, may be related to the comparatively high original content of organic matter in the soil.

Ferguson and Gorby (1971) reported appreciable losses of soil N, ranging in magnitude from 12 to 24% over a 12-year period (1953–1965) in soil subjected to various wheat–fallow and forage rotations. These results differ somewhat from those reported by Biederbeck et al. (1984) and Freyman et al. (1982) who observed no appreciable losses or even net gains in soil N in the Brown and Dark Brown soil zones, respectively. However, wetter conditions in the Black soil at Brandon usually results in significant N losses from leaching and denitrification. It is also difficult to increase the inherently high content of organic matter in Black soils.

Transitional and Gray Luvisolic soil zones

The influence of fallowing frequency on measurements of soil quality in the Gray soil zones has not been documented. Campbell (unpublished), however, found evidence of significant short-term fluctuations in concentrations of organic matter within a 3-year F–W–W rotation in a Dark Gray soil at Somme, Sask. Concentrations were highest prior to the fallow phase and lowest after the fallow phase. Similar fluctuations were evident in a 6-year rotation including 1 year of fallow, 3 years of wheat, and 2 years of hay. No consistent difference was observed in concentrations of organic matter between the 3- and 6-year rotations.

Solonetzic soils

On a solonetzic soil at Vegreville, Alta., Carter (1984) found that, in accordance with observations in other soil zones, soil concentrations of organic C and N 22 years after treatment commenced were appreciably higher in the continuous wheat treatment than in a F–W treatment (Table 26). Further, the soil's potential capacity to mineralize N (N_0) was 73% higher in the continuous wheat treatment. Biological activity, measured as microbial biomass C, mineral N flush, or soil respiration, was consistently higher in the continuous wheat treatment than in the F–W rotation.

Continuous cropping also improved physical soil properties, as evident from measurements of extractable sodium concentrations, water infiltration rate, breaking strength of the Bnt horizon, shrinkage of the Bnt horizon, and aggregate stability measured by wet sieving.

Carter (1984) also observed that the inclusion of alfalfa–bromegrass forage in an 8-year rotation with cereals (5 years forage and 3 years cereals) significantly enhanced concentrations of organic C and N relative to those in monoculture wheat rotations (Table 19). Indices of biological activity in the soil of the cereal–forage rotation were comparable to those of the continuous wheat treatment. Physical properties, notably extractable sodium, infiltration rate, shrinkage of the Bnt horizon, and aggregate stability, were generally comparable between these two treatments.

ECONOMIC PERFORMANCE

Although producers ought to choose crop rotations that ensure the long-term sustainability and competitiveness of agricultural production, they are also obliged to emphasize short-term economic viability.

Expected net income

A primary criterion in the evaluation of any crop rotation is the level of net income that producers can reasonably expect to achieve from adopting it, given their particular set of physical factors, economic considerations, and personal characteristics. Producers generally choose the crop rotation that provides the greatest expected net income.

Expected net income levels have been computed for several of Agriculture Canada's crop rotation experiments in western Canada (Zentner et al. 1984*b*, Zentner et al. 1986, Zentner and Campbell 1988, Zentner et al. 1988*b*). These values, expressed on the basis of a unit of cultivated land and of 1984–1986 input cost conditions, represent the income remaining after paying for all cash costs, depreciation on buildings and machinery, and salary for farm labor. They represent the funds available for income tax, principal payments on farm debt, and interest allowance on owned equity.

Brown soil zone

At low wheat prices all rotations at Swift Current lost money, but F–W lost the least (Fig. 20) (Zentner et al. 1984*b*, Zentner and Campbell 1988). At medium wheat prices, well-fertilized F–W–W was the most profitable rotation; only at high wheat prices did continuous wheat prove to be the most profitable rotation.

Net income for rotations that included flax, fall rye, or oat hay generally ranked low relative to those that included only wheat. Substituting flax for wheat was profitable only when the price for flax was greater than 2.4 times the price for wheat (Zentner and Campbell 1988). Similarly, replacing wheat with fall rye on fallow in a 3-year rotation was profitable only when the rye-to-wheat price ratio exceeded 0.85.

Application of N and P fertilizers was profitable for continuous wheat under most economic situations and often minimized losses (Fig. 21). However, for fallow-containing wheat rotations only when either wheat prices were high or fertilizer costs low did the application of both N and P fertilizers produce higher net incomes than for rotations to which only N or P fertilizer was applied.

Dark Brown soil zone

At Scott, as at Swift Current, fertilized F–W was most profitable at low wheat prices, F–W–W at intermediate prices, and continuous wheat at high wheat prices (Fig. 20) (Zentner et al. 1986). When the canola-to-wheat price ratio was greater than 2.5, the substitution of canola for wheat grown on fallow in the 3-year rotation became profitable. Rotations that included alfalfa hay crops generally

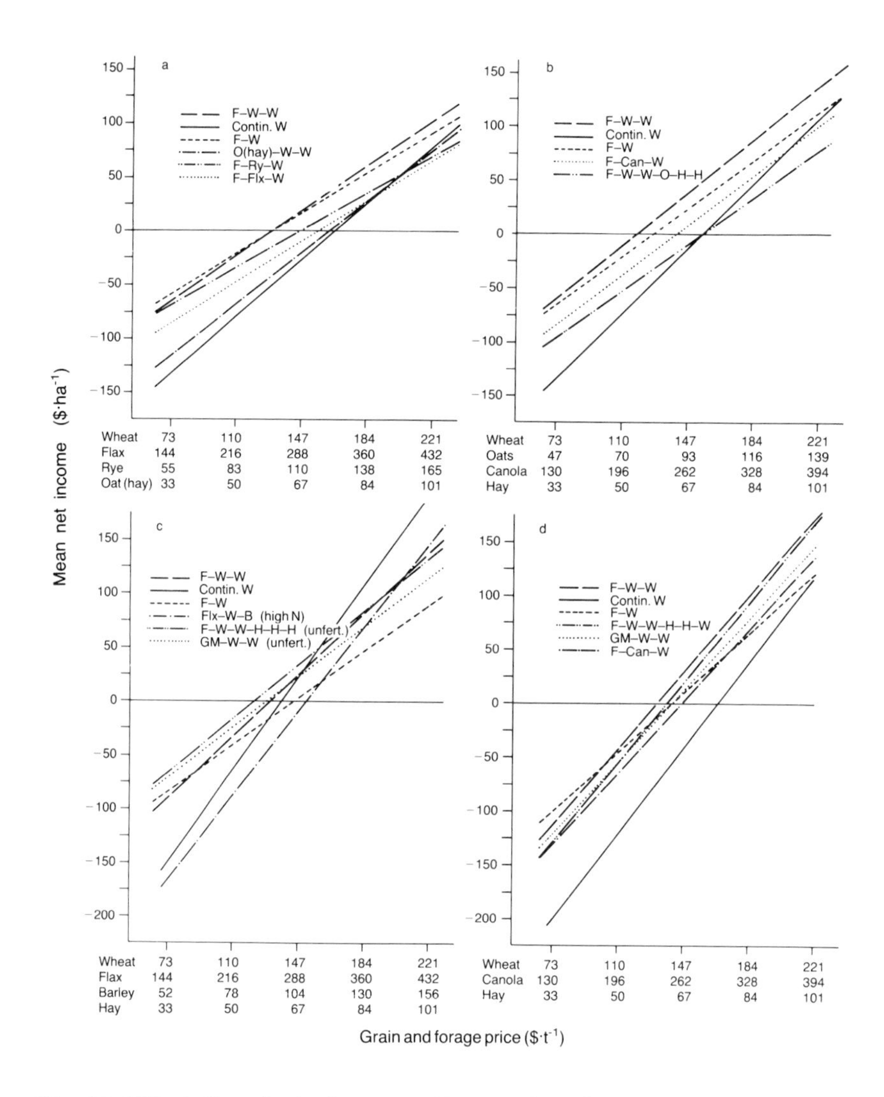

Fig. 20 Effect of product prices on net income for selected rotations in Saskatchewan at (*a*) Swift Current (1967–1984), (*b*) Scott (1972–1984), (*c*) Indian Head (1978–1984), and (*d*) Melfort (1972–1984) (redrawn from Zentner et al. 1986).

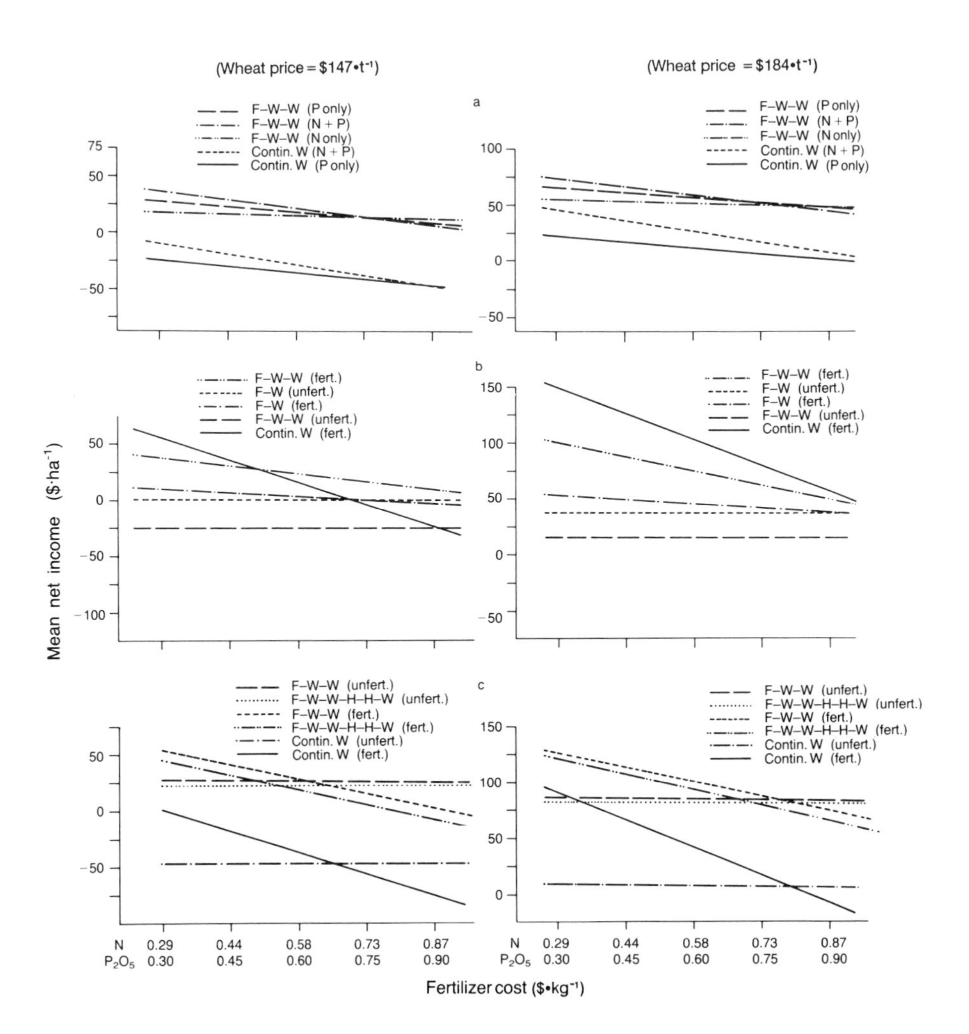

Fig. 21 Effect of wheat price and fertilizer cost on net income from fertilization for (*a*) Swift Current (1967–1984), (*b*) Indian Head (1978–1984), and (*c*) Melfort (1972–1984) (redrawn from Zentner et al. 1986).

provided low net income, reflecting difficulties in establishing alfalfa in dry stubble land.

An economic analysis of six rotations that included grain sorghum over a 6-year period at Lethbridge (Janzen et al. 1987) showed that gross economic returns (yield × prices in 1984) were similar for all except F–S rotations in which they were significantly lower. Yields of wheat and sorghum were comparable, but sorghum commanded a lower price and required greater management; consequently, it does not appear to be a profitable alternative for southern Alberta at present.

Black soil zone

At Indian Head, four rotations, fertilized F–W–W and continuous wheat, and unfertilized GM–W–W and F–W–W–H–H–H, showed good economic performance under most reasonable economic situations (Zentner et al. 1986, Zentner et al. 1988*b*). During 1960–1977 when fertilized treatments received low rates of N and P based on use of the general recommendations for the region, the unfertilized GM–W–W provided the highest net income for wheat prices ranging from $73 to $221 per tonne; unfertilized F–W–W–H–H–H generally ranked second highest (Zentner et al. 1988*b*). During 1978–1984 when fertilizer was applied at higher rates based on soil tests, fertilized continuous wheat was the most profitable rotation at wheat prices greater than $180 per tonne (Fig. 20). When wheat prices were between $145 and $180 per tonne, the net incomes produced by fertilized F–W–W and continuous wheat, and unfertilized F–W–W–H–H–H were not only similar but also the highest. At wheat prices below $145 per tonne net income was highest (or losses were lowest) for the unfertilized legume-containing rotations. Fertilized F–W ranked in the three rotations earning highest net incomes when grain prices were low.

The economic benefit from fertilization at Indian Head was significant in most years, particularly for the more intensive crop rotations and in years of favorable moisture (Fig. 21). Proper fertilization had the favorable effect of amplifying the levels of net income in years of favorable growing conditions and diminishing the economic losses in years of unfavorable growing conditions. There were indications that the fertilizer rates used in the early periods of the study were lower than the economic optimums (Zentner et al. 1988*b*). Reducing the cost of fertilizer while holding the quantities used constant increased the mean economic returns of the rotations receiving fertilizer; the relative effects were greatest for the continuous cropping rotations and lowest for the rotations containing fallow. Summer fallow acts as a short-term substitute for fertilizer because of nitrogen mineralization during this period. Low fertilizer costs also reduced the economic value of the complementary effect of including forage and legume crops as green manure in the rotation.

At Brandon, Spratt et al. (1975) studied six crop rotations that included various fallow substitute crops and found little economic value in including a high proportion of summer fallow in the rotation, especially on sandy loam soil (Table 29). Potatoes proved to be highly profitable on clay loam and sandy loam soils. Despite occasional failure, sweetclover hay, corn, oat hay, and flax also proved to be equally or more profitable than summer-fallowing once every 3rd year.

At Melfort, when wheat prices were greater than $110 per tonne, the most profitable rotation was fertilized F–W–W (Fig. 20) (Zentner et al. 1986). Fertilized F–W–W–H–H–W ranked second highest at higher product price levels, whereas at lower prices mean net incomes differed little among the fertilized F–W, F–W–W, GM–W–W, or F–W–W–H–H–W rotations. Net income for fertilized continuous wheat ranked lowest at all grain prices because of problems with weeds and disease (e.g., root rot). Substituting canola for wheat grown on fallow in a 3-year rotation was profitable only when the canola-to-wheat price ratio exceeded 2.0. At lower fertilizer costs or when grain prices were high, or both, N and P fertilization increased net income (Fig. 21). The economic benefit from fertilization was lowest for the F–W–W–H–H–W rotation likely reflecting an improvement in soil quality produced by the forage–legume crop.

The results at Melfort differed somewhat from those observed for comparable rotations over the same period at Indian Head. Among the fertilized monoculture wheat rotations, F–W–W was highly profitable at both locations; continuous wheat ranked high at Indian Head, but lowest at Melfort, whereas the mean net incomes obtained for fertilized F–W were similar at both locations. The economic benefit from fertilization was also considerably lower at Melfort than at Indian Head. Further, net returns of fertilized GM–W–W at Melfort were often lower than for a comparable monoculture of fertilized

Table 29 Mean net income for six crop rotations at Brandon, Man., 1965–1970

Rotation sequence	Clay loam soil† ($·ha^{-1})	Sandy loam soil† ($·ha^{-1})
F–W–W	45	10
SC–W–W	55	20
Po–W–W	173	101
Co–W–W	45	17
O(hay)–W–W	49	17
Flx–W–W	42	12

Source: Spratt et al. 1975.
† Based on 1970 prices and costs.

wheat and were also lower than for a comparable rotation that neither received fertilizer nor included a legume crop as green manure. These results are also opposite to those at Indian Head where an unfertilized rotation including a legume as green manure often provided the highest net income. Rotations that included grass–legume systems generally produced good economic returns at both locations. These differences in economic performance of the rotations among locations were attributed to differences in the physical and biological characteristics of the soils, weather, weed and pest problems, and the differing amounts of N and P fertilizer applied at the two sites (Zentner et al. 1986).

Costs of production

Total cash costs per unit of cultivated land are shown in Table 30, and average cash cost per unit of wheat produced is given in Table 31.

Table 30 Total cash costs for crop rotations in the Brown, Dark Brown, and Black soil zones of Saskatchewan

Rotation sequence	Fertilizer		Cost ($\$\cdot ha^{-1}$)	% of control
	N	P		
Swift Current (1967–1984)†				
F–W (control)	√	√	83	100
F–W–W	√	√	104	125
F–W–W	0	√	100	120
F–W–W	√	0	93	112
F–Flx–W	√	√	111	134
F–Ry–W	√	√	94	113
Contin. W	√	√	187	225
Contin. W	0	√	166	200
Flx–W–W	√	√	181	218
O(hay)–W–W	√	√	181	218
Scott (1972–1984)††				
F–W (control)	√	√	85	100
F–C	√	√	76	89
F–W–W	√	√	112	132
F–C–W	√	√	105	124
F–C–O	√	√	103	121
F–W–O–H	√	√	102	120
F–W–W–O–H–H	√	√	108	127
Contin. W	√	√	172	202

(*continued*)

Table 30 Total cash costs for crop rotations in the Brown, Dark Brown, and Black soil zones of Saskatchewan *(concluded)*

Rotation sequence	Fertilizer N	Fertilizer P	Cost ($\$\cdot ha^{-1}$)	% of control
Indian Head (1978–1984)‡				
F–W (control)	√	√	89	100
F–W	0	0	80	90
F–W–W	√	√	133	149
F–W–W‡‡	√	√	149	167
F–W–W	0	0	101	113
GM–W–W	0	0	86	97
F–W–W–H (4 years)	0	0	100	112
F–W–CL–O–H (4 years)	0	0	84	94
F–W–W–H–H–H	0	0	83	93
Contin. W	√	√	240	270
Contin. W	0	0	156	175
Flx–W–B	√	√	207	233
Flx–W–B (high N)	√	√	228	256
Flx–W–B	0	0	143	161
Melfort (1972–1984)††				
F–W (control)	√	√	121	100
F–W–W	√	√	163	134
F–W–W	0	0	115	94
F–C–W	√	√	156	129
GM–W–W	√	√	160	132
F–C–W–W§	√	√	176	145
F–W–W–H–H–W	√	√	183	151
F–W–W–H–H–W	0	0	125	103
F–C–W–H–H–C	√	√	178	146
Contin. W	√	√	243	200
Contin. W	0	0	157	130

† From Zentner and Campbell (1988).
†† From Zentner et al. (1986).
‡ From Zentner et al. (1988*b*).
‡‡ Straw was baled and sold.
§ 1977–1984 only.

Table 31 Average cash cost per unit of wheat produced at four locations in Saskatchewan

Rotation sequence	Fertilizer N	Fertilizer P	Cost ($·t−1)	% of control
Swift Current (1967–1984)				
F–W (control)	√	√	88	100
F–W–W	√	√	94	107
F–W–W	0	√	95	108
F–W–W	√	0	96	109
Contin. W	√	√	140	159
Contin. W	0	√	144	164
Scott (1972–1984)				
F–W (control)	√	√	65	100
F–W–W	√	√	70	108
Contin. W	√	√	93	144
Indian Head (1978–1984)				
F–W (control)	√	√	77	100
F–W	0	0	75	97
F–W–W	√	√	86	112
F–W–W	0	0	98	127
GM–W–W	0	0	68	88
Contin. W	√	√	105	136
Contin. W	0	0	171	222
Melfort (1972–1984)				
F–W (control)	√	√	80	100
F–W–W	√	√	84	105
F–W–W	0	0	74	92
GM–W–W	√	√	91	114
Contin. W	√	√	118	147
Contin. W	0	0	111	138

Brown soil zone

At Swift Current, total cash costs were highest for N and P fertilized continuous wheat, intermediate for fertilized F–W–W, and lowest for F–W. Substituting flax or oat hay for wheat had little effect on total cash expenditures. However, substituting fall rye for wheat grown on fallow in the 3-year rotation reduced cash costs by $10 per hectare as a result of lower requirements for herbicide and fewer tillage operations during summer fallow; with fall rye the fallow period is reduced from 21 to 14 months. The addition of recommended rates of N and P fertilizer increased total cash cost per unit of land but had little effect on the average cost of producing wheat (Table 31).

Dark Brown soil zone

At Scott, results were generally similar to Swift Current. Total cash costs and average cost of producing wheat were lowest for the 2-year F–W rotations, highest for continuous wheat, and intermediate for F–W–W (Tables 30 and 31). Total cash costs for the 3-year rotations that included canola or those that included forage crops were similar to those for F–W–W.

Black soil zone

On Black soils, total cash costs increased with cropping intensity and the use of N and P fertilizers (Table 30). At Indian Head, total cash expenditures were lowest for F–W, unfertilized GM–W–W, and for two of the three legume–forage rotations; at Melfort they were lowest for F–W, and the unfertilized F–W–W and F–W–W–H–H–W. Total cash costs for continuous wheat were similar at both locations, but those for F–W and F–W–W were generally higher at Melfort than at Indian Head because much higher rates of fertilizer were used after 1972 at Melfort. In contrast, the average cost per unit of wheat produced was generally similar in comparable fertilized monocultures of wheat at both locations (Table 31). At Indian Head, fertilization reduced the average unit cost of producing wheat in the more intensive rotations, whereas at Melfort, a possible excess of N applied to fallow-seeded crops after 1972 produced the opposite effect.

In all soil zones the resource categories most affected by increases in the rotation length or changes in the cropping mix were fertilizers, pesticides, labor, and machine operation (fuel and oil plus machine repair) (Fig. 22). Within the fertilized monocultures of wheat, fertilizer costs showed the greatest increase with cropping intensity reflecting the higher rates of N applied to stubble-seeded crops. Costs for herbicides and insecticides also increased with the more intensive crop rotations, but they were generally lower for rotations that included fall rye, forage, or legume crops as green manure because little herbicide was applied to these crops and to the cereal areas being undersown to forage or sweetclover.

Fuel and oil expenses increased with rotation length, but to a lesser extent than other inputs because the additional fuel required for planting and harvesting the extra crop was partially offset by the savings from reduced tillage of summer fallow. Expenditures on machine repairs also increased with intensity of cropping, particularly for planting and harvesting equipment.

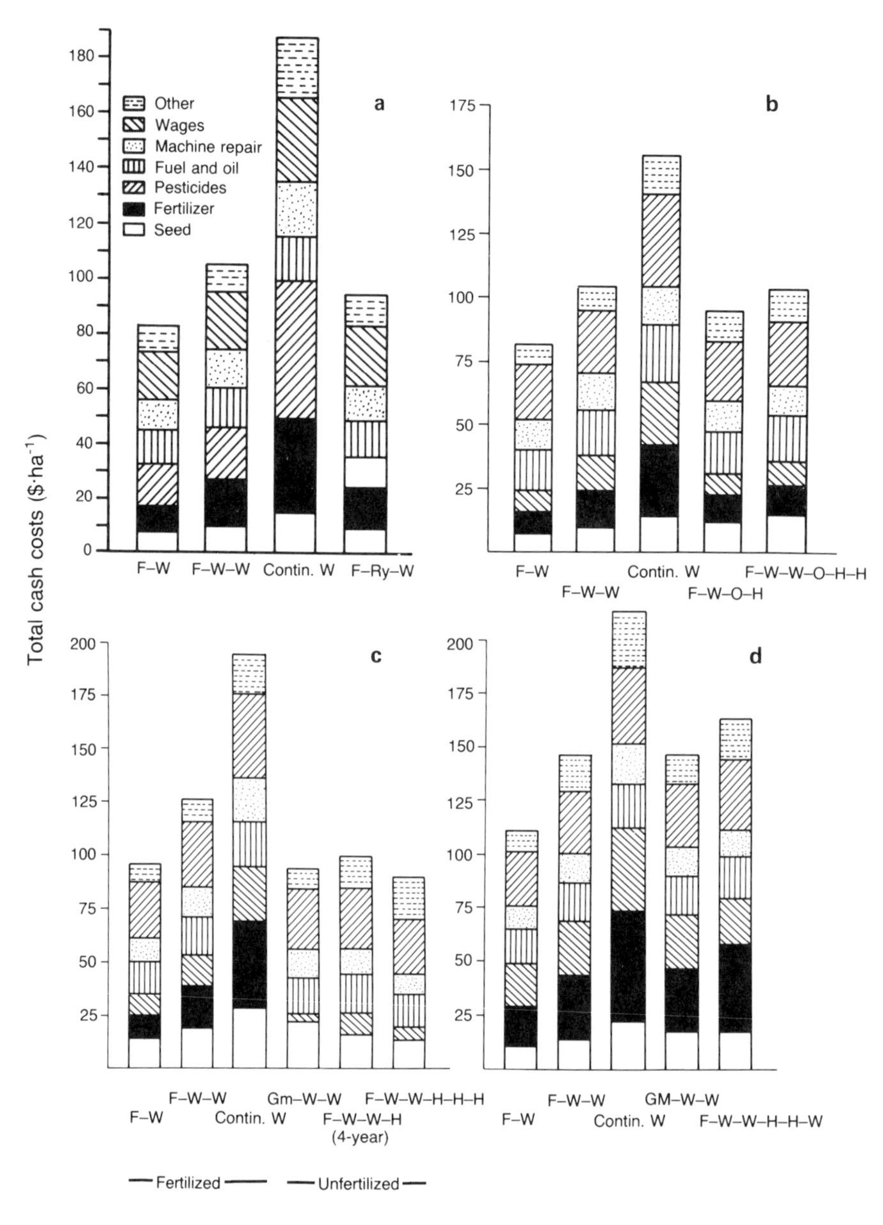

Fig. 22 Total cash costs by resource category for selected rotations in Saskatchewan at (*a*) Swift Current, (*b*) Scott, (*c*) Indian Head, and (*d*) Melfort (redrawn from Zentner et al. 1986).

The direct relationship between labor requirements and cropping intensity reflects the greater use of tillage, planting, and harvesting equipment. Furthermore, the seasonality of the labor requirements was greatly affected by the rotation length and cropping mix. Under the F–W rotation, total labor requirements were spread uniformly over the growing season (Zentner et al. 1984*b*); however, as rotations were lengthened, labor requirements became concentrated in the spring and fall. Further, crop rotations that include fall rye or forage crops have greater requirements for summer labor as compared with rotations that include only spring-sown crops.

Seed costs and other cash costs, such as for building repair, interest on operating capital, and land taxes, generally increased in direct proportion to the area being cropped. Capital expenditures, in the form of additional grain storage and planting and harvesting machinery, were also greater with the continuous-type rotations.

Breakeven yield ratios

The breakeven stubble-to-fallow yield ratios for wheat represent the average relationship that must exist between the yields of wheat grown on stubble and on fallow so that equivalent levels of net income are obtained (Table 32). When actual stubble-to-fallow yield ratios

Table 32 Estimated and actual breakeven stubble-to-fallow yield ratios for fertilized wheat rotations† at four locations in Saskatchewan

	Assumed wheat price ($\$\cdot t^{-1}$)				Actual yield ratio
Rotation	110	147	184	221	
Swift Current (1967–1984)					
F–W–W	0.81	0.75	0.70	0.67	0.74
Contin. W	0.95	0.87	0.79	0.74	0.71
Scott (1972–1984)					
F–W–W	0.75	0.70	0.66	0.63	0.81
Contin. W	0.78	0.73	0.69	0.66	0.64
Indian Head (1978–1984)					
F–W–W	0.97	0.85	0.78	0.73	0.93
Contin. W	0.99	0.89	0.81	0.76	0.94
Melfort (1972–1984)					
F–W–W	0.88	0.78	0.73	0.69	0.91
Contin. W	0.87	0.78	0.72	0.68	0.67

† Based on total cash costs only.

are greater than the breakeven values, then stubble cropping of wheat is more profitable than fallow cropping, and vice versa.

Breakeven values vary considerably with the expected price for wheat. In general, stubble cropping was profitable for the F–W–W rotation at all locations. In contrast, at most locations, continuous cropping was usually profitable only when grain prices were high. It was profitable at all price scenarios at Indian Head where moisture is rarely limiting and where weed and disease problems can be kept under control.

Inclusion in the breakeven calculations of other considerations, such as overhead costs or government policies (e.g., grain delivery quotas), increases the breakeven stubble-to-fallow yield ratios.

Income variability

Decision making in a risky environment usually entails making a trade-off between increases in expected net income and increases in income variability. A higher level of expected net income generally requires giving up a lower level of income variability (Fig. 23). Low standard deviations indicate low income variability or risk. Producers who are averse to risk (i.e., do not like to gamble) are less willing to adopt crop rotations with high income variability. In general, rotations that included high proportions of summer fallow or forage crops had the lowest income variability and would therefore be favored by producers who are highly averse to risk (Fig. 23). Continuous-type rotations generally had the highest income variability and would be favored by those producers who are less averse to risk. As shown, all-risk crop insurance is effective in reducing, but not eliminating, the additional risk associated with the extended crop rotations.

Brown soil zone

At Swift Current, F–W had the lowest income variability; Flx–W–W and the continuous wheat rotations had the highest (Fig. 23). Further, the extent and frequency of economic losses increased with cropping intensity but decreased with grain price. For example, when wheat price was $147 per tonne, economic losses were incurred in 5, 8, and 9 of 18 years for fertilized F–W, F–W–W, and continuous wheat, respectively; however, when the price for wheat was $184 per tonne, economic losses were incurred in only 1, 2, and 5 of 18 years in these same rotations. Consequently, producers who are averse to risk will likely only consider using a F–W or F–W–W rotation in this region. From a purely economic standpoint, fertilized continuous wheat would likely only be considered with all-risk crop insurance, when marketing opportunities and product prices are expected to be very high, and only by those producers with low aversion to risk.

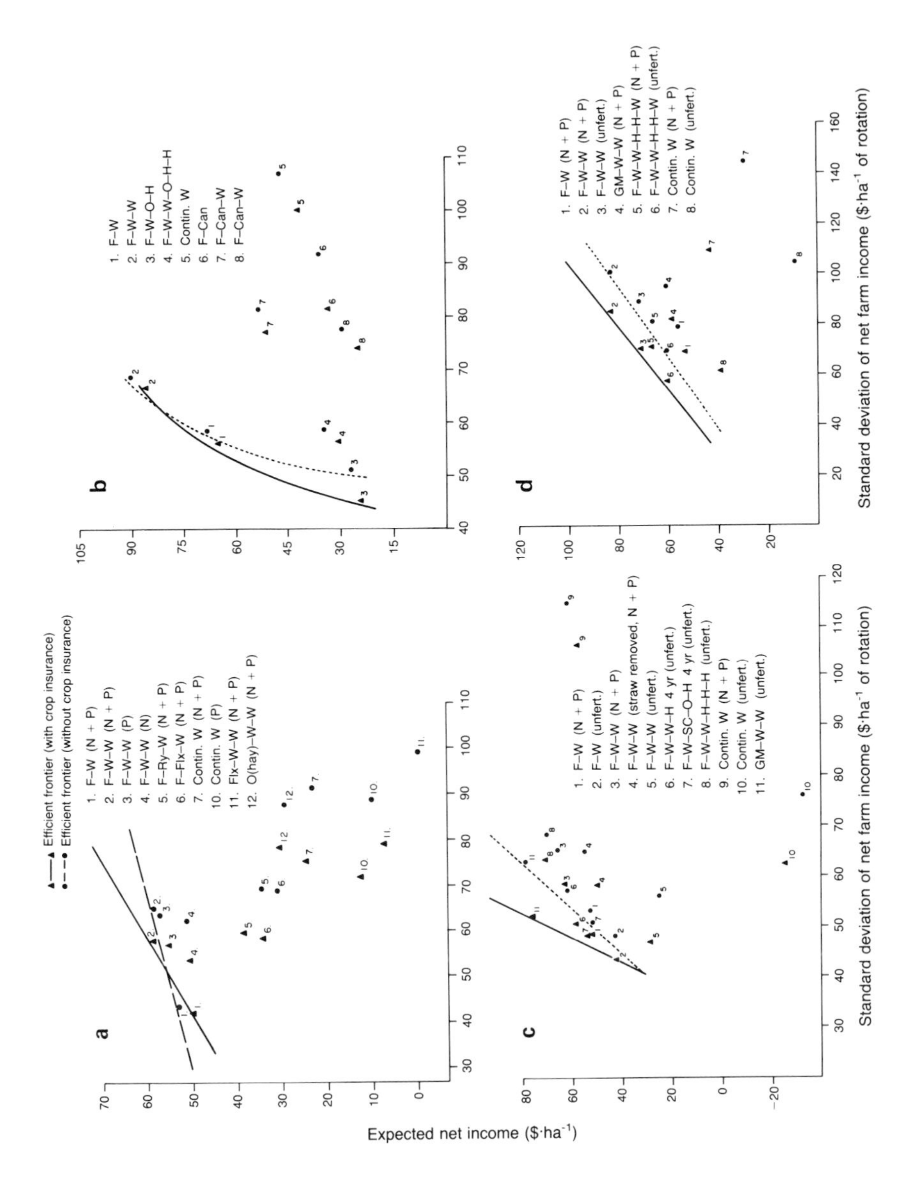

Fig. 23 Trade-off between making a profit and income variability and the effect of the crop insurance program on this relationship at (*a*) Swift Current, (*b*) Scott, (*c*) Indian Head, and (*d*) Melfort (redrawn from Zentner et al. 1986).

An alternative method of assessing the risks of stubble cropping was developed for the Brown soil zone (Zentner et al. 1987*a*). This method uses a formula relating yields of wheat grown on fertilized stubble land to depth of moist soil at time of planting and level of growing season (May–July) rainfall. The equation is a modification of equation 1 in the section on "Soil moisture":

$$\text{yield}\ (\text{kg}\cdot\text{ha}^{-1}) = [-424 + 11.7 \times \text{depth of moist soil at planting (cm)} + 90.9 \times \text{growing season rainfall (cm)}] \quad [4]$$

The depth of moist soil refers to the depth to which a soil probe (Brown et al. 1981) can be pushed manually into the soil in early spring. This value, together with knowledge of the soil's texture, can be used to estimate available moisture in the soil.

The relationship shown in equation 4, together with the historical distribution of growing season rainfall received at Swift Current were used to compute the probability of obtaining a yield of stubble-seeded wheat that was high enough to allow recovery of total cash costs for soils with various depths of moist soil in early spring.

Table 33 shows that the yields of stubble-seeded wheat needed to recover cash costs of $125 and $175 per hectare decreased when expected wheat price increased from $90 to $120 per tonne.

Using the yield equation and the historical growing season rainfall data, the probabilities of achieving the breakeven yields are shown in Table 34. The results indicate that, if the soil is moist to only 30 cm at planting and the expected price for wheat is $90 per tonne, then the probability of obtaining a yield for stubble-seeded wheat sufficient to cover cash costs of $125 per hectare (or the equivalent yield of 1389 $\text{kg}\cdot\text{ha}^{-1}$) is 38%. Similarly, the probability of recovering cash costs of $175 per hectare (equivalent yield of 1944 $\text{kg}\cdot\text{ha}^{-1}$) is only 1.5 out of 10 years (15% probability) at this low price for wheat. Higher prices for wheat, or greater levels of soil moisture in spring improved the chance of recovering input costs.

Table 33 Breakeven yields of stubble-seeded wheat

Expected wheat price ($\$\cdot\text{t}^{-1}$)	Yield needed to recover cash costs of $\$125\cdot\text{ha}^{-1}$ ($\text{kg}\cdot\text{ha}^{-1}$)	Yield needed to recover cash costs of $\$175\cdot\text{ha}^{-1}$ ($\text{kg}\cdot\text{ha}^{-1}$)
90	1389	1944
100	1250	1750
110	1136	1591
120	1042	1458

Table 34 Chances of at least breaking even with stubble-seeded wheat at Swift Current, Sask.

Depth of moist soil (cm)	Expected wheat price ($·t^{-1})			
	90	100	110	120
	(years out of 10)			
Recover cash costs of $125·ha^{-1}				
30	3.8	4.6	5.6	5.8
45	5.0	5.6	6.5	7.4
60	6.3	7.3	7.9	8.4
75	7.9	8.3	8.8	9.0
Recover cash costs of $175·ha^{-1}				
30	1.5	2.0	2.5	3.3
45	2.1	2.8	3.4	4.5
60	2.9	3.7	4.7	5.8
75	3.8	4.9	5.7	6.7

Dark Brown soil zone

At Scott (Fig. 23), income variability was lowest for F–W–O–H and highest for continuous wheat. All rotations produced economic losses in some years, the extent and frequency of which were greatest for continuous wheat (32% of the years when wheat price was $184 per tonne), intermediate for F–W–W (16% of the years), and lowest for F–W (11% of the years).

When product prices are low, producers in the Dark Brown soil zone who are highly averse to risk would likely choose the F–W–W or F–W rotation. When product prices are higher and with crop insurance, these same producers would likely choose only the F–W–W rotation. Conversely, producers with low risk aversion would choose F–W–W when product prices are low, and F–W–W or continuous wheat when product prices are high. Rotations including legume-forage crops would not likely be chosen by most producers (given poor performance of these rotations in this experiment), whereas rotations including canola (primarily F–Can–W) would be chosen during periods when oilseed prices are relatively high or when marketing opportunities for cereal crops are poor.

Black soil zone

In the Black soil zone at Indian Head producers who expect product prices to be low and are highly averse to risk would be most willing to choose among unfertilized GM–W–W, or F–W–W–H–H–H, and the fertilized F–W–W and F–W rotations (Fig. 23). When product prices are expected to be high, highly risk averse producers would likely choose from among unfertilized GM–W–W, F–W–W–H (4 years), F–W–W–H–H–H, and fertilized F–W–W; producers who are somewhat less averse to risk would also choose from among fertilized Flx–W–B, and continuous wheat.

At Melfort, when product prices are low, risk-averse producers would likely choose fertilized F–W–W or F–W–W–H–H–W (Fig. 23). Producers who are less averse to risk might also choose F–Can–W or continuous wheat if product prices were expected to be high.

ENERGY CONSIDERATIONS

Technological developments that lower unit costs of production, competition for labor from industrial sectors, and the drudgery associated with many farm tasks have prompted the mechanization of western Canadian agriculture. This substitution of capital for labor has increased the dependency of agriculture on nonrenewable forms of energy (Gayton 1982, Stirling 1979). The dependency, combined with rising costs and possible shortages of energy in the future, has created a need to consider the use of more energy efficient production practices.

Brown soil zone

Analysis of energy efficiencies of the various crop rotation studies have rarely been done. In the Brown soil zone at Swift Current, total nonrenewable energy input per hectare of land was lowest for the traditional F–W rotation, intermediate for N and P fertilized F–W–W, and highest for N and P fertilized continuous wheat. Thus the more intensive cropping systems are highly vulnerable to rising energy costs unless product prices respond similarly (Table 35) (Zentner et al. 1984*a*, Zentner et al. 1989). Substituting flax or rye for wheat in the rotations reduced the total input of energy by 3–8%, whereas withholding fertilizer reduced the total input of energy by 16–37%. Liquid fuel (for field operations and local product transport) and fertilizer (primarily N) were the major nonrenewable energy inputs; both increased with cropping intensity (Zentner et al. 1989). Fuel accounted for 30–50% of the total energy input of the rotations; fertilizer represented 15–49% and was of greater importance than fuel for the continuous crop rotations. Despite the high energy content in pesticides, they accounted for only 4–11% of the total energy input of the rotations because of their low rates of application.

Table 35 Mean annual energy input, energy output, and energy efficiency of rotations at Swift Current, Sask., 1967–1984

Rotation sequence	Fertilizer		Total energy input		Metabolizable energy output†		Output:input ratio	Wheat produced per unit of energy input (kg·GJ −1)
	N	P	Mean (MJ·ha −1)	Y_{cv} (%)	Mean (MJ·ha −1)	Y_{cv} (%)		
F–W	√	√	3482	10	12639	31	3.6	262
F–W–W	√	√	4470	14	14641	34	3.3	240
F–W–W	0	√	3787	11	14015	35	3.7	268
F–W–W	√	0	3741	16	13088	36	3.6	261
F–Flx–W	√	√	4317	17	8374	40	2.0	–
F–Ry–W	√	√	4200	15	14119	43	3.4	–
Contin. W	√	√	7100	23	17764	44	2.6	191
Contin. W	0	√	4472	20	15124	41	3.4	245
Flx–W–W	√	√	6517	23	12331	43	2.0	–
O(hay)–W–W	√	√	7162	22	12603	42	1.8	–
$S_{\bar{x}}$†			62		178		0.05	5

Source: Zentner et al. 1989.

† Metabolic energy output was calculated as the quantity of grain produced (less seed requirements) multiplied by the metabolic energy content of the grain for human consumption.

The output of metabolizable energy for human consumption increased with cropping intensity, reflecting the higher total annual production of grain (Table 35). In contrast, the energy output-to-input ratios and the quantity of wheat produced per unit of energy input (energy efficiency) decreased with cropping intensity. Rotations that included flax or cereal forage crops had the lowest energy efficiencies.

Other soil zones

The effect on the energy inputs of replacing summer fallow with a legume crop for green manure, for typical rotations in the Brown, Dark Brown, Black, and Gray Luvisol soil zones of Alberta, has been examined by Rice and Biederbeck (1983) (Table 36). The calculations are based on the use of an annual legume such as Tangier flatpea for the Brown and Dark Brown zones, with available N increased by 100 $kg \cdot ha^{-1}$ through N_2 fixation. For the Black and Gray Luvisol soil zones, the calculations were based on the use of a clover crop underseeded in the second barley crop, providing an additional 60 and 125 $kg \cdot ha^{-1}$ of available N for the Black and Gray Luvisol soils, respectively. The use of legume crops as green manure has the potential for net energy conservation of 500–2850 $MJ \cdot ha^{-1}$. In terms of diesel equivalents (1.4–7.3 $L \cdot ha^{-1}$) and fertilizer N equivalents (supplying N at 5.4–24.3 $kg \cdot ha^{-1}$) this could provide very substantial relief for the tightening squeeze between production costs and product price.

Table 36 Calculated energy inputs for cropping rotations with and without legumes in the four soil zones in Alberta

Inputs	Energy inputs ($MJ \cdot ha^{-1}$)							
	Brown		Dark Brown		Black		Gray Luvisol	
	F–W–F–W	GM–W–F–W	F–W–B	GM–W–B	F–C–B–B	GM–C–B–B	F–C–B–B	GM–C–B–B
Machinery production	414.2	414.2	506.4	506.4	844.9	844.9	844.9	944.9
Machinery maintenance	301.0	301.0	365.2	365.2	609.4	609.4	609.4	609.4
Gasoline	911.8	911.8	1357.15	1309.0	756.4	756.4	869.9	869.9
Diesel fuel	729.8	673.1	1011.6	758.4	1941.9	1773.2	2232.4	2038.7
Nitrogen	652.1	54.2	1882.1	674.4	3040.6	1390.9	5045.9	2588.1
Phosphorus	71.2	122.2	115.3	183.2	298.9	298.9	315.6	315.6
Seed	845.5	966.8	1418.9	1580.5	1070.3	1389.2	1070.3	1389.2
Herbicide	17.8	17.8	24.8	24.8	569.4	53.4	569.4	53.4
Labor	3.2	3.2	6.5	6.5	5.0	5.0	6.7	6.7
Total	3946.6	3464.3	6688.3	5408.4	9136.8	7121.3	11564.5	8715.9
Energy conservation from legumes in rotation								
Net energy ($MJ \cdot ha^{-1}$)	482.3		1279.9		2015.5		2848.6	
Fuel ($MJ \cdot ha^{-1}$)	56.7		301.7		168.7		193.4	
Diesel equivalent ($L \cdot ha^{-1}$)	1.4		7.3		4.1		4.7	
Fertilizer ($MJ \cdot ha^{-1}$)	546.9		1139.8		1649.7		2457.8	
Fertilizer-N equivalent ($kg \cdot ha^{-1}$)	5.4		11.3		16.3		24.3	

Source: Rice and Biederbeck 1983.

DISCUSSION

CURRENT STATUS OF ROTATIONS

The results of the rotation studies are such that the discussion can be divided into two parts: first, the drier region that covers the Brown and Dark Brown soils; and second, the wetter region that covers the Black and Gray Luvisolic soils.

Brown and Dark Brown soils

The uncertainty in amount and distribution of moisture received in this region have limited the choice of crops used in rotations to drought-hardy spring wheat and summer fallow, although oat hay, flax, fall rye, clovers (for green manure), and, more recently, canola have been included. Of the rotations tested, F–W–W provided the highest economic benefits; it was more risky than F–W but contributed less to soil degradation. Continuous wheat, even when properly fertilized, provided the lowest economic returns, but it was the best for soil conservation. Substitution of oat hay, flax, canola, or fall rye for spring wheat did not greatly improve the economic returns (though this depends on prices). Fall rye provided an advantage in soil conservation by reducing leaching of NO_3-N and by improving soil organic matter. Winter wheat and newly developed high-yielding Canada prairie spring wheats have not been used for a sufficient time in rotations to allow proper assessment of their potential. Lentils and sweetclover for green manure have not performed well when they were used because of lack of management expertise (especially in control of weeds). In this region, perennial crops such as alfalfa and grasses offer little advantage in cereal rotations, because their great moisture use affects subsequent cereal crops deleteriously for two or three seasons after plow down.

Producers usually adopt systems that have the potential to provide economic benefits, such as the shorter, fallow-containing rotations that accelerate soil degradation. Consequently, some other cropping system that has the potential to provide an adequate economic return and minimize soil degradation is required for this drier region. Perhaps a rotation that includes 1 year of fallow followed by 3, 4, 5, or 6 years of wheat would be ideal. This sequence would reduce soil degradation caused by frequent fallowing yet would allow eradication of grassy weeds by cheaper tillage methods rather than by use of costly herbicides. Unfortunately, few studies of this intermediate length have been carried out. The system may be improved by adopting a conservation tillage approach and by fertilizing according to soil test recommendations. The benefits of fertilization for increasing yields, economic returns, and improving soil quality were demonstrated in studies at Swift Current and

Lethbridge. Increasing the current cutting height of the crops, from 15 cm at harvest to heights recommended by new technology such as "snow trapping," enhances the storage of soil moisture and may reduce the risk of soil moisture limiting the frequency of stubble cropping. The success of this approach is supported by recent studies at Scott (Brandt and Kirkland 1986) and Swift Current (Zentner et al. 1988*a*). The Scott study demonstrated the independent benefits of fertilizing, snow trapping, and herbicide use, with a complementary effect on crop yields when all three agronomic practices were combined (Table 37).

Another potential improvement that could be adopted in crop rotations in this drier zone is the inclusion of annual legumes for green manure (e.g., Black lentils) (Slinkard et al. 1987), which might reduce fertilizer costs. Other studies have demonstrated that, if managed properly (e.g., seed early, seed shallow into chemical fallow or stubble), winter wheat may be a reasonable substitute for 1 or 2 years of spring wheat in a rotation (Economic Regional Development Agreement projects in progress). Furthermore, this crop may be better suited to take advantage of the moisture distribution in the drier areas than spring wheat. However, winter wheat has low protein content and has the added risk of being susceptible to winterkill, rust, and snow mold. New wheat varieties, such as the

Table 37 Effect of management practices on average yields of wheat grown continuously over 3 years (1983–1985) at Scott (Dark Brown soil zone) and Kindersley (Brown soil zone), Sask.

Herbicide and fertilizer treatments	Stubble treatment: Fall tillage flattened stubble ($kg \cdot ha^{-1}$)	Stubble treatment: Snow trap tall stubble ($kg \cdot ha^{-1}$)
No herbicide, no fertilizer (control)	855	962
Diclofop-methyl plus chlorsulfuron or bromoxynil, no fertilizer	1134	1187
No herbicide, N at 90 $kg \cdot ha^{-1}$, P_2O_5 at 45 $kg \cdot ha^{-1}$	1200	1380
Diclofop-methyl plus chlorsulfuron or bromoxynil, N at 90 $kg \cdot ha^{-1}$, P_2O_5 at 45 $kg \cdot ha^{-1}$	1444	1758

Source: Brandt and Kirkland 1986.

Canada prairie spring wheats, may prove useful substitutes for the traditional hard red spring or durum wheats, but their short straw and susceptibility to bunt and loose smut has so far not encouraged their use in this region. Barley was rarely used in any of the rotation studies in this region, perhaps because the area is too dry to guarantee malt quality (low protein) or high yields (feed barley). Grain sorghum was used in a rotation study at Lethbridge, but it had no advantage over spring wheat. Corn for grain or silage was not included in any study.

In the future, if wheat prices remain low, many new and exotic crops probably will be used, but it is unlikely that any of them will displace spring wheat as the number one component of the crop rotation in the drier area. Summer fallow continues to play a significant role in the region because of unpredictable moisture. However, its frequency of use will decrease as management reduces the use of mechanical energy in favor of herbicides for weed control.

Black and Gray Luvisolic soils

In the Black and Gray soil zones summer-fallowing to store moisture had no beneficial effect on crop yields. The economics of continuous cropping, though superior to similar systems in the drier region, was less promising than the shorter F–W–W, GM–W–W, or the mixed F–cereal–forage systems. Continuous cropping, especially when properly fertilized, enhanced soil quality. However, legume-containing systems were found to enhance soil quality to a greater extent than most grain rotations; they also appeared to reduce the incidence of weeds and some injurious insects. Unfortunately, legume–cereal rotations are not used by many producers because they require more sophisticated management skills, are more labor-intensive, and require more varied equipment. On the other hand, producers who use these rotations have a choice either of feeding hay or of selling it to other producers or to the dehydration industry, or they may plow it down, thereby improving soil quality and productivity.

Canola and barley are often part of the rotation. These crops, when properly managed, are no more deleterious to soil quality than wheat; they are good alternatives as cash crops, and canola may help to break cycles of some insect pests in wheat and, in turn, benefit from a suppression of soilborne diseases by the wheat. Pulse crops, such as peas and lentils, have been used to replace summer fallow, especially as herbicides have become available to control weeds. Winter wheat production should be feasible in areas that receive adequate snow cover to ensure safe overwintering of the crop. However, winterkill and snow mold continue to pose a real problem. Furthermore, the crop must be rotated with a crop that is early maturing (e.g., barley) so that the winter wheat can be seeded in mid to late August. Late seeding may extend maturation time and increase the chances of winterkill and disease (e.g., rust). Winter wheat also offers the advantage that, because it is harvested early, it is able to escape early frosts, which are common to this region. The new high-yielding Canada prairie spring wheats may also provide a good alternative crop. The short straw is an advantage in this area where excess straw is a problem with lodging, straw incorporation, and weed control. In this region, the need to apply adequate amounts of fertilizers to improve the efficiency of water use was demonstrated by the rotation studies. High input of N and P fertilizers for increasing yields and crop quality and for maintaining soil quality are required. Recent development of equipment to seed into heavy trash and standing stubble, banding fertilizers into soil below crop residue layers, and availability of more effective herbicides should encourage continuous cropping with conservation tillage.

QUO VADIS?

Where do we go from here? There is no doubt that the array of crop rotations now in place have provided a wealth of knowledge to the scientific community and to producers. However, these rotations remain a valuable reservoir of soil and agronomic information that has not been fully exploited by agricultural scientists. To date, within Agriculture Canada, the Swift Current and Lethbridge studies have permitted in-depth analysis of soil nutrient dynamics. These studies have also yielded valuable information on economics and crop management.

What does the future hold for crop rotation studies within Agriculture Canada? An efficient way to obtain answers to such questions is to use an approach such as that proposed by Janzen (1986). He described three types of rotation studies (Fig. 24): fundamental, feasibility, and reference experiments.

Fundamental experiments are designed to elucidate mechanisms and processes that underlie a potential crop rotation system. Such studies would likely be conducted under controlled conditions in laboratory, growth chamber, or field. They can serve to identify and define key processes in a soil–crop system and thus can allow early identification of superior rotational components that could make more efficient use of available soil, crop, and climatic resources. These experiments can help to reduce the number of treatments that need to be rigorously tested in full-scale field experiments. They can also facilitate the development of models that can be used to predict long-term effects of a given rotation on soil productivity and thus, in some instances, reduce the need for many long-term rotation studies. A number of fundamental studies continue at Lethbridge and Swift Current. In one such study, ^{15}N-labeled legume material is used in microplots to determine which is the most suitable annual legume for use as green manure in the Brown and Dark Brown soils.

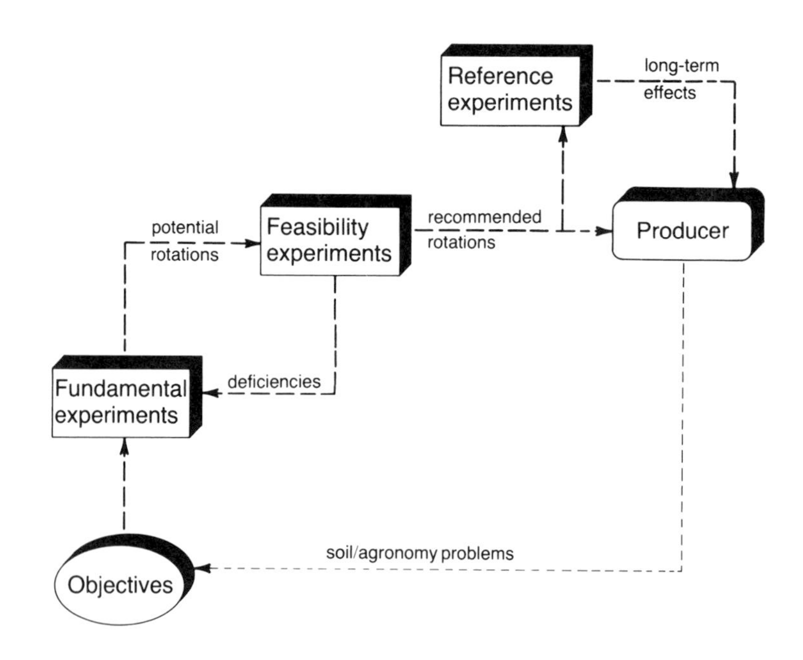

Fig. 24 Proposed strategy for future crop rotation research (redrawn from Janzen 1986).

Feasibility studies are used to sort out the more appropriate rotations or crops to be used in a long-term study. Assessments are based on economic and agronomic viability; they continue for a period of 5 to 6 years and require that detailed data be collected on yields and crop quality, susceptibility to pests and to soil degradation, soil fertility, soil biology, and required economic and energy assessment. Feasibility of various crop sequences and possibilities of allelopathic effects may also be tested. The advantage of this type of experiment is that it provides data that are more currently applicable as it promotes a faster evolution of improved rotations by reducing generation time. Preferably the study should be carried out at several diverse sites to compensate for its limited duration. An example of this type of experiment is the sorghum rotation study carried out at Lethbridge. Other studies that fit this category include studies recently initiated at Brandon, Indian Head, and Vegreville. At Brandon they are examining the interaction of tillage (zero versus minimum) and crop rotations (Flx–W–Co–B and Soy–W–Co–B) on crop productivity and weed dynamics (R.D. Dryden and C.A. Grant, personal communications). At Indian Head the effect of three, 4-year rotations (F–W–W–WW, W–W–Flx–WW, and W–Flx–WW–Pe) on productivity, weeds, and insect dynamics are being studied (G.P. LaFond, personal communication). At Vegreville a study initiated in 1988 involves five barley rotations (B–B–F, B–B–Pe, B–B–GM, B–B–C, and B–B–O) on a Black, Dark Brown, and Gray Luvisol, which are monitored intensively for their effect on crop productivity and soil quality; ^{15}N is used to determine N dynamics. Other examples are the snow management study of Brandt and Kirkland (1986) mentioned earlier, and an experiment located at several western research stations that is designed to identify the most promising annual legume crops for green manure (Slinkard et al. 1987).

Reference experiments are used to monitor long-term (decades) trends in crop and soil productivity. Here soil changes should be determined at regular intervals of 5–10 years, but agronomic data should be collected annually. Because of the high need for resources, number of treatments and replications are limited. These rotations usually lack flexibility and often yield results that are dated by progressive changes in agronomic practices. Nonetheless, these studies are invaluable for testing system models and for providing data that cannot be obtained in any other way. Most of the major continuing rotation studies at Agriculture Canada research stations in western Canada fit into this category.

At Swift Current a new, long-term, crop rotation study was initiated in 1986 to answer some questions raised by producers that were not answered by earlier studies. The study includes intermediate-length rotations, flexible cropping based on weed infestation or level of spring soil moisture, use of annual legumes as green manure, and crops such as winter wheat and high-yielding Canada prairie spring wheat (HY320). The study uses currently recommended agronomic technology (e.g., fertilizer placement and minimum tillage). Through feasibility studies involving tillage, snow management, winter wheat, and Canada prairie spring wheats, and feasibility and fundamental studies with legumes for green manure, the rotation was designed to produce scientific and technological information.

The schematic presented in Fig. 24 depicts the type of unified research strategy that future rotation studies might wish to employ. The scheme provides for effective and efficient collection of data and minimizes the stress on available resources.

CONCLUSIONS

Summer fallow remains a legitimate option in the cropping systems of western Canada, though its role and recommended use vary depending on edaphic and climatic conditions. In the Black and Gray soil zones, where moisture deficits are relatively small, summer-fallowing can be justified only for control of otherwise unmanageable pests or in the event of potential drought. In the Brown and Dark Brown soils, however, where moisture stress is the primary yield-limiting factor, the replenishment of soil moisture during fallow reduces economic risks and warrants the inclusion of some summer fallow in crop rotations. Although frequent inclusion of fallow enhances soil degradation, this effect can be minimized by using techniques for conservation tillage, by use of partial fallows (e.g., green manure or cereal hay crops), or by reducing the frequency of fallow in the rotation. Economic analyses indicate that the optimum frequency for using fallow in the Brown and Dark Brown soil zones is about once in 3 years, though that value varies depending on soil and economic variables.

Inclusion of perennial forages in crop rotations represents an important means of improving soil quality, crop nutrition, and crop yields in the Black, Dark Gray, and Gray soil zones. Economic analyses indicate that extended rotations of spring wheat, with several years of forage grown for hay, generate favorable net economic returns. In the Brown and Dark Brown soil zones, the inclusion of perennial forages is not recommended because of excessive depletion of soil moisture by these deep-rooted crops.

The influence of one crop on a subsequent crop in the rotation is largely contingent on use of soil moisture by the first crop, residual fertility effects, and effect on pest populations. Oilseed crops (flax and canola) generally deplete soil moisture to a lesser extent than cereal crops and thereby increase the potential yields of subsequent crops in drier soils. These same crops, however, particularly flax, do not compete well with weeds and may therefore suppress the yield of subsequent crops by allowing weeds to proliferate. Regardless of cropping system, periodic rotation of crops is recommended for the control of certain weeds, diseases, and insects.

Fertilizers are assuming progressively greater importance in cropping systems as indigenous soil fertility declines and cropping systems are intensified. Appropriate application of fertilizers generally increases expected net returns, except in rotations having high frequency of fallow or forage and when costs of fertilizers are high. Aside from directly increasing yields, fertilizer application has three effects. It increases the efficiency of moisture use by stimulating root growth; it improves moisture conservation and snow-trapping by increasing surface crop residue, and it enhances long-term soil productivity by increasing the content and quality of organic matter.

Ammonium-based fertilizers may depress soil pH, though significant acidification will likely occur only over the long term in most soils.

Long-term crop rotations indicate that agronomic practices greatly affect soil quality. They demonstrate, further, that soil productivity can be maintained indefinitely by adoption of economically viable, conservation-oriented management practices. Although most soils have exhibited an inevitable decline in concentrations of organic matter following initial cultivation, this trend has been halted or even reversed by the use of appropriate crop sequences and fertilizer strategies. Crop yields have generally increased over the decades since inception of arable agriculture, though the relative contributions of technological advances and soil quality dynamics to this trend are uncertain. Results of these studies therefore suggest that the dire predictions of inevitable declines in soil productivity are not necessarily a fait accompli, at least not for all producers.

The ideal cropping system, in most conditions, retains considerable flexibility for crop selection in response to dynamics in soil moisture reserves, economic variables, and infestations of weeds or other pests. Although most rotation experiments have adopted rigid cropping sequences to facilitate the interpretation of data, the results can be interpreted to assist in the design of crop sequences with sufficient flexibility to exploit changes in economic and agronomic conditions. Although most producers have long-term general cropping strategies, short-term revisions in response to economic and climatic factors are both anticipated and recommended.

Decisions regarding cropping strategies should take into consideration not only short-term benefits but also their long-term effects on soil and environmental quality. Crop rotation studies have demonstrated that agronomic practices exert strong influence on the concentration of organic matter, soil erodibility, soil pH, soil biota activity, and various other indices of soil quality. Less well known, but of equal importance, are the effects of various cropping practices on the environment through mechanisms such as nitrate leaching, carbon dioxide evolution, groundwater contamination with pesticides, and accumulation of pesticides in soil and farm produce. All these ramifications deserve consideration in the design of optimum cropping systems.

Interpretation and extrapolation of crop rotation results is meaningful only in light of the limitations inherent in the rotation experiments. Foremost among these is the limited number of sites at which the rotation studies were conducted. Furthermore, in most cases, the experiments were located on relatively productive soils of uniform topography and medium texture. Other potential limitations, which need to be taken into consideration in the interpretation of the results, include the limited number of crops examined, lack of replication in some long-term studies, and possible confounding effects of gradual changes in cultivars and management practices over time. Because of evolving agronomic practices, currently recommended management systems generally have not been subjected to long-term experimental consideration.

The crop rotation studies summarized in this publication have identified a number of agronomic facets that deserve more detailed examination or that can be improved upon in future studies. Of particular importance is the paucity of information regarding the influence of cropping practices on weed, disease, and insect infestation. As well, details of soil conditions have been examined only in a few experiments. These deficiencies demonstrate the need for a comprehensive, team-oriented approach to crop rotation research. Cropping practices or components that deserve particular emphasis in the future include organic or sustainable production systems; flexible cropping systems, which reduce economic risk without resorting to frequent fallowing; partial fallow systems such as green manuring; and more diverse cropping systems using a greater variety of crop species.

The findings reported in this publication will benefit both the agricultural extension and scientific communities. Results presented represent the basis for the formulation of agronomic strategies which could be disseminated to producers. Some of the extension information currently propagated is based on anecdotal or testimonial evidence. Scientific evidence in this publication may confirm or invalidate this information or, furthermore, spawn additional agronomic recommendations. For the scientific community, this publication not only will provide background information, but also will serve to identify areas requiring research attention. As well, these studies may represent opportunities for the development or verification of simulation models, and in-depth, critical examinations of specific mechanisms.

REFERENCES

Anderson, C.H. 1971. Comparison of tillage and chemical summerfallow in a semiarid region. Can. J. Soil Sci. 51:397–403.

Anderson, D.T. 1961. Surface trash conservation with tillage machines. Can. J. Soil Sci. 41:99–114.

Anderson, D.T. 1967. The cultivation of wheat. Pages 338–355 *in* Nielsen, K.F., ed. Canadian Centennial Wheat Symposium. Western Coop. Fertilizers Ltd., Calgary, Alta.

Anderson, D.W.; Gregorich, E.G. 1984. Effect of soil erosion on soil quality and productivity. Pages 105–113 *in* Soil erosion and land degradation, Proceedings 2nd Annual Western Provinces Conference Rationalization of Water and Soil Research and Management, Saskatoon, Sask.

Austenson, H.M. 1978. Principles of agronomy with particular reference to Saskatchewan conditions. University of Saskatchewan, Saskatoon, Sask. 70 pp.

Ball, W.S. 1987. Crop rotations for North Dakota. NDSU Extension Service, N.D. State University Bull. EB-48. 20 pp.

Biederbeck, V.O. 1988. Replacing fallow with annual legumes for plow-down or feed. Pages 46–61 *in* Symposium on Crop Diversification in Sustainable Agriculture Systems, 27 Feb. 1988. University of Saskatchewan, Saskatoon, Sask.

Biederbeck, V.O.; Campbell, C.A. 1987. Effect of wheat rotations and fertilization on soil microorganisms and enzymes in a Brown loam. Pages 153–164 *in* Proceedings Soils and Crops Workshop, February 1987. University of Saskatchewan, Saskatoon, Sask.

Biederbeck, V.O.; Campbell, C.A.; Schnitzer, M. 1986. Effect of wheat rotations and fertilization on microorganisms and biochemical properties of a Brown loam in Saskatchewan. Pages 552–553 *in* Transactions 13th Congress of the International Society of Soil Science (vol. II). Hamburg, Aug. 1986.

Biederbeck, V.O.; Campbell, C.A.: Smith, A.E. 1987. Effects of long-term 2,4-D field applications on soil biochemical processes. J. Environ. Qual. 16:257–262.

Biederbeck, V.O.; Campbell, C.A.; Zentner, R.P. 1984. Effect of crop rotation and fertilization on some biological properties of a loam in southwestern Saskatchewan. Can. J. Soil Sci. 64:355–367.

Bowren, K. 1977. The effect of reducing tillage on summerfallow when weeds are controlled with herbicides in the Black soil zone of Saskatchewan. Pages 5–11 *in* Proceedings Soil Fertility and Crops Workshop, University of Saskatchewan, Saskatoon, Sask.

Bowren, K.E. 1984. Crop rotation studies. Pages 87–93 *in* Research Report 1984. Research Station, Agriculture Canada, Melfort, Sask.

Bowren, K.E.; Biederbeck, V.O.; Bjorge, H.A.; Brandt, S.A.; Goplen, B.P.; Henry, J.L.; Ukrainetz, H.; Wright, T.; McLean, L.A. 1986. Soil improvement with legumes. Sask. Dep. Agric. Publ. M10-86-02. 24 pp.

Bowren, K.E.; Cooke, D.A. 1975. Effects of legumes in cropping systems in northeastern Saskatchewan. Can. J. Plant Sci. 55:351.

Bowren, K.E.; Cooke, D.A.; Downey, R.K. 1969. Yield of dry matter and nitrogen from top and roots of sweetclover, alfalfa and red clover at five stages of growth. Can. J. Plant Sci. 49:61–69.

Bowren, K.E.; Dryden, R.D. 1971. Effect of fall and spring treatment of stubble land on yield of wheat in the Black soil region of Manitoba and Saskatchewan. Can. Agric. Eng. 13:32–35.

Bowren, K.E.; Townley-Smith, L. 1986. Review of tillage and crop rotation studies, Melfort. Pages 4–18 *in* Agriculture Canada Research Branch Work Planning Meeting on Tillage and Rotations. 16 July 1986. Regina, Sask.

Brandt, S.A. 1981. Cropping sequences on Gray-Wooded soils. Pages 146–154 *in* Proceedings Soils and Crops Workshop. February 1981. University of Saskatchewan, Saskatoon, Sask.

Brandt, S.A. 1984. Crop rotation studies at Scott. Page 39 *in* Research highlights 1984. Saskatoon Research Station, Saskatoon, Sask.

Brandt, S.A.; Keys, C.H. 1982. Effect of crop rotations on soil moisture levels. Pages 38–48 *in* Proceedings Soils and Crops Workshop. February 1982. University of Saskatchewan, Saskatoon, Sask.

Brandt, S.A.; Kirkland, K.J. 1986. Wheat rotations for the Brown and Dark Brown soil zones. Pages 227–237 *in* Slinkard, A.E.; Fowler, D.B., eds. Wheat production in Canada—A review. Proceedings Canadian Wheat Production Symposium. 3–5 March 1986. Saskatoon, Sask.

Brown, Paul L. 1964. Legumes and grasses in dryland cropping systems in the northern and central great plains—A review of literature. U.S. Department of Agriculture, USDA-ARS Misc. Publ. No. 952. Washington, D.C. 64 pp.

Brown, Paul L.; Black, A.L.; Smith, Charles M.; Enz, John W.; Caprio, J.M. 1981. Soil water guidelines and precipitation probabilities for barley and spring wheat in flexible cropping systems in Montana and North Dakota. Montana Coop. Ext. Serv. Bull. 356. 30 pp.

Campbell, C.A.; de Jong, R.; Zentner, R.P. 1984*a*. Effect of cropping, summerfallow and fertilizer nitrogen on nitrate-nitrogen lost by leaching on a Brown Chernozemic loam. Can. J. Soil Sci. 64:61–74.

Campbell, C.A.; Read, D.W.L.; Biederbeck, V.O.; Winkleman, G.E. 1983*a*. The first 12 years of a long-term crop rotation study in southwestern Saskatchewan—nitrate N distribution in soil and N uptake by the plant. Can. J. Plant Sci. 63:563–578.

Campbell, C.A.; Read, D.W.L.; Zentner, R.P.; Leyshon, A.J.; Ferguson, W.S. 1983*b*. First 12 years of a long-term crop rotation study in southwestern Saskatchewan—Yield and quality of grain. Can. J. Plant Sci. 63:91–108.

Campbell, C.A.; Read, D.W.L.; Winkleman, G.E.; McAndrew, D.W. 1984*b*. First 12 years of a long-term crop rotation study in southwestern Saskatchewan—bicarbonate-P distribution in soil and P uptake by the plant. Can. J. Soil Sci. 64:125–137.

Campbell, C.A.; Zentner, R.P. 1984. Effect of fertilizer on soil pH after 17 years of continuous cropping in southwestern Saskatchewan. Can. J. Soil Sci. 62:651–656.

Campbell, C.A.; Zentner, R.P.; Dormaar, J.F.; Voroney, R.P. 1986. Land quality, trends and wheat production in western Canada. Pages 318–353 *in* Slinkard, A.E.; Fowler, D.B., eds. Wheat production in Canada. University of Saskatchewan, Saskatoon, Sask.

Campbell, C.A.; Zentner, R.P.; Johnson, P.J. 1988*a*. Effect of crop rotation and fertilization on the quantitative relationship between spring wheat yield and moisture use in southwestern Saskatchewan. Can. J. Soil Sci. 68:1–16.

Campbell, C.A.; Zentner, R.P.; Selles, F. 1988*b*. Regressions for estimating straw yields and N and P contents of spring wheat and N mineralization in a Brown loam soil. Can. J. Soil Sci. 68:337–344.

Campbell, C.A.; Zentner, R.P.; Steppuhn. H. 1987. Effect of crop rotations and fertilizers on moisture conserved and moisture use by spring wheat in southwestern Saskatchewan. Can. J. Soil Sci. 67:457–472.

Carder, A.C.; Hennig, A.M.F. 1966. Soil moisture regimes under summerfallow, wheat and red fescue in the upper Peace River region. Agr. Meteorol. 3:311–331.

Carter, M.R. 1984. Effect of soil management on some chemical, physical and biological properties of a solonetzic soil. Soil Sci. 138:411–416.

Chinn, S.F.F.; Ledingham, R.J. 1958. Application of a new laboratory method for the determination of the survival of Helminthosporium sativum spores in soil. Can. J. Bot. 36:289–295.

Cook, R.J. 1981. The influence of rotation crops on take-all decline phenomenon. Phytopathology 71:189–192.

Coote, D.R.; Dumanski, J.; Ramsey, J.F. 1981. An assessment of the degradation of agricultural land in Canada. Land Resource Research Institute, Contrib. No. 188, Ottawa.

Dormaar, J.F. 1983. Chemical properties of soil and water-stable aggregates after 67 years of cropping to spring wheat. Plant Soil 75:51–61.

Dormaar, J.F. 1984. Monosaccharides in hydrolysates of water-stable aggregates after 67 years of cropping to spring wheat as determined by capillary gas chromatography. Can. J. Soil Sci. 64:647–656.

Dormaar, J.F.; Pittman, U.J. 1980. Decomposition of organic residues as affected by various dryland spring wheat–fallow rotations. Can. J. Soil Sci. 60:97–106.

Dubetz, S. 1983. Ten-year irrigated rotation U. 1911–1980. Agric. Can. Tech. Bull. 1983-21E. 11 pp.

Ferguson, W.S.; Gorby. B.J. 1971. Effect of various periods of seed-down to alfalfa and bromegrass on soil nitrogen. Can. J. Soil Sci. 51:65–73.

Freyman, S.; Palmer, C.J.; Hobbs, E.H.; Dormaar, J.F.; Schaalje, G.B.; Moyer, J.R. 1982. Yield trends in long-term dryland rotations at Lethbridge. Can. J. Plant Sci. 62:609–619.

Gayton, D.V. 1982. Direct energy use and conservation potential on Saskatchewan straight grain farms. Pages 687–690, *in* Conference Proceedings, Energex 1982, vol. II, Solar Energy Society of Canada, Winnipeg, Man.

Hennig, A.M.F.; Rice, W.A. 1977. Effects of date of sod breaking on nitrogen requirements and yield of barley following fescue in rotation. Can. J. Soil Sci. 57:477–485.

Hermans, J.C. 1977. An update of research on solonetzic soils. Pages 173–178 *in* Soil conservation reclamation and research. Proceedings Alberta Soil Science Workshop. 1–2 February 1977. Edmonton, Alta.

Hopkins, E.S.; Barnes, S. 1928. Crop rotation and soil management for the Prairie Provinces. Can. Dep. Agric. (n.s.) Bull. 98, 53 pp.

Hopkins, E.S.; Leahey, A. 1944. Crop rotations in the Prairie Provinces. Can. Dep. Agric. Farmers' Bull. 124, 70 pp.

Hoyt, P.B.; Hennig, A.M.F. 1971. Effect of alfalfa and grasses on yield of subsequent wheat crops and some chemical properties of a Gray-Wooded soil. Can. J. Soil Sci. 51:177–183.

Hoyt, P.B.; Leitch, R.H. 1983. Effects of forage legume species on soil moisture, nitrogen and yield of succeeding barley crops. Can. J. Soil Sci. 63:125–136.

Hoyt, P.B.; Rice, W.A.; Hennig, A.M.F. 1978. Utilization of northern Canadian soils for agriculture. Proceedings 11th Congress International Society Soil Science 3:333–346.

Janzen, H.H. 1986. Rotation research at Lethbridge. Pages 61–89 *in* Agriculture Canada Research Branch Work Planning Meeting on Tillage and Rotations. 16 July 1986. Regina, Sask.

Janzen, H.H. 1987*a*. Effect of fertilizer on soil productivity in long-term spring wheat rotations. Can. J. Soil Sci. 67:165–174.

Janzen, H.H. 1987*b*. Soil organic matter characteristics after long-term cropping to various spring wheat rotations. Can. J. Soil Sci. 67:845–856.

Janzen, H.H.; Major, D.J.; Lindwall, C.W. 1987. Comparison of crop rotations for sorghum production in southern Alberta. Can. J. Plant Sci. 67:385–393.

Janzen, P.J.; Korven, H.A.; Harris, G.K.; Lehane, J.J. 1960. Influence on depth of moist soil at seeding time and seasonal rainfall on wheat yields in southeastern Saskatchewan. Agriculture Canada, Ottawa, Ont. Publ. 1090. 10 pp.

Kirkland, K.J.; Keys, C.H. 1981. The effect of snow trapping and cropping sequence on moisture conservation and utilization in west-central Saskatchewan. Can. J. Plant Sci. 61:241–246.

Kraft, D.F. 1980. Adoption of technology and crop production. *In* Proceedings Prairie Production Symposium: Soils and Land Resources. Advisory Committee to the Canadian Wheat Board, University of Saskatchewan, Saskatoon, Sask.

Lal, R.; Steppuhn, H. 1980. Minimizing fall tillage on the Canadian prairies—A review. Can. Agric. Eng. 22:101–106.

Ledingham, R.J. 1961. Crop rotations and common root rot in wheat. Can. J. Plant Sci. 41:479–486.

Lindwall, C.W.; Anderson, D.T. 1981. Agronomic evaluation of minimum tillage systems for summerfallow in southern Alberta. Can. J. Plant Sci. 61:247–253.

Luke, H.H.; Pfahler, P.L.; Barnett, R.D. 1983. Control of Septoria nodorum on wheat with crop rotation and seed treatment. Plant Dis. 67:949–951.

Matthews, G.D. 1949. Progress report, 1937–1947. Dominion Experimental Station, Swift Current, Sask. Canada Department of Agriculture, Ottawa, Ont.

McAndrew, D.W. 1986. Classical rotation and tillage research at Vegreville and Lacombe. Pages 51–56 *in* Agriculture Canada

Research Branch Work Planning Meeting on Tillage and Rotations. 16 July 1986. Regina, Sask.

McGill, W.B.; Campbell, C.A.; Dormaar, J.F.; Paul, E.A.; Anderson, D.W. 1981. Soil organic matter losses. Pages 72–133 *in* Agricultural land: Our disappearing heritage—A symposium. Proceedings 18th Annual Alberta Soil Science Workshop, Edmonton, Alta.

Michalyna, W.; Hedlin, R.A. 1961. A study of moisture storage and nitrate accumulation in soil related to wheat yields on four cropping sequences. Can. J. Soil Sci. 41:5–15.

Molberg, E.S.; McCurdy, E.V.; Wenhardt, A.; Dew, D.A.; Dryden, R.D. 1967. Minimum tillage requirements for summerfallow in western Canada. Can. J. Soil Sci. 47:211–216.

Morrall, R.A.A.; Dueck, J. 1982. Epidemiology of sclerotinia stem rot of rapeseed in Saskatchewan. Can. J. Plant Pathol. 4:161–168.

National Soil Erosion–Soil Productivity Research Planning Committee. 1981. Soil erosion effects on soil productivity: A research perspective. J. Soil Water Conserv. 36:82–90.

Nuttall, W.F.; Bowren, K.E.; Campbell, C.A. 1986. Crop residue management practices, and N and P fertilizer effects on crop response and on some physical and chemical properties of a Black Chernozem over 25 years in a continuous wheat rotation. Can. J. Soil Sci. 66:159–171.

O'Halloran, I.P. 1986. Phosphorus transformations in soils as affected by management. Ph.D. Thesis, University of Saskatchewan, Saskatoon, Sask.

Piening, L.J.; Orr, D. 1988. Effects of crop rotation on common root rot of barley. Can. J. Plant Pathol. 10:61–65.

Pittman, U.J. 1977. Crop yields and soil fertility as affected by dryland rotations in southern Alberta. Comm. Soil Sci. Plant Anal. 8:391–405.

Poyser, E.A.; Hedlin, R.A.; Ridley, A.O. 1957. The effect of farm and green manure on the fertility of blackearth-meadow clay soils. Can. J. Soil Sci. 37:48–56.

Prairie Farm Rehabilitation Administration. 1982. Land degradation and soil conservation issues on the Canadian prairies: An overview. Prairie Farm Rehabilitation Administration, Soil and Water Conservation Branch.

Rennie, D.A.; Ellis, J.G. 1978. The shape of Saskatchewan. University of Saskatchewan, Saskatoon, Sask. Saskatchewan Institute of Pedology Publ. M41. 61 pp.

Rice, W.A.; Biederbeck, V.O. 1983. The role of legumes in the maintenance of soil fertility. Pages 35–42 *in* Proceedings 20th Annual

Alberta Soil Science Workshop, 22–23 February 1983. Edmonton, Alta.

Rice, W.; Hoyt, P. 1980. Crop rotations the role of legumes: Alfalfa production in the Peace River region. Northern Research Group 80-2:E1–E8.

Ripley, P.O. 1969. Crop rotation and productivity. Agriculture Canada, Ottawa, Ont. Publ. 1376, 78 pp.

Ritchie, J.T. 1981. Soil water availability. Plant Soil 58:327–338.

Robertson, J.A. 1979. Lessons from the Breton plots. University of Alberta. Agric. For. Bull. 2:8–13.

Robertson, J.A.; McGill, W.B. 1983. New directions for the Breton plots. University of Alberta. Agric. For. Bull. 6:41–45.

Russell, R.C. 1934. Studies of take-all and its causal organism, *Ophiobolus graminis* Sacc. Dom. Can. Dep. Agric. Bull. 170, N.S. 64 pp.

Shaner, G. 1981. Effect of environment on fungal leaf blight of small grains. Ann. Rev. Phytopathol. 19:273–296.

Shipton, W.A.; Boyd, W.R.J.; Rosielle, A.A.; Shearer, B.I. 1971. The common Septoria diseases of wheat. Bot. Rev. 37:231–262.

Slinkard, A.E.; Biederbeck, V.O.; Bailey, L.; Olson, P.; Rice, W.; Townley-Smith, L. 1987. Annual legumes as a fallow substitute in the northern Great Plains of Canada. Pages 6–7 *in* Power, J.F., ed. The role of legumes in conservation tillage systems. Proceedings Soil Conservation Society America Conference. University of Georgia, Athens, Ga., 27–29 April 1987.

Sparrow, Hon. H.O. 1984. Soil at risk—Canada's eroding future. A report on soil conservation by the standing committee on agriculture, fisheries and forestry, to the Senate of Canada. 129 pp.

Spratt, E.D. 1966. Fertility of a chernozemic clay soil after 50 years of cropping with and without forage crops in the rotation. Can. J. Soil Sci. 46:207–212.

Spratt, E.D.; Strain, J.H.; Gorby, B.J. 1975. Summerfallow substitutes for western Manitoba. Can. J. Plant Sci. 59:685–689.

Stirling, B. 1979. Use of non-renewable energy on Saskatchewan farms. Saskatchewan Science Council Background Study 2. 38 pp.

Staple, W.J.; Lehane, J.J. 1952. The conservation of soil moisture in southern Saskatchewan. Sci. Agric. 32:36–47.

Staple, W.J.; Lehane, J.J. 1954. Wheat yield and use of moisture on substations in southern Saskatchewan. Can. J. Agric. Sci. 34:460–468.

Staple, W.H.; Lehane, J.J.; Wenhardt, A.V. 1960. Conservation of soil moisture from fall and winter precipitation. Can. J. Soil Sci. 40:80–88.

Stewart, J.W.B.; McKenzie, R.H.; Moir, J.O.; Dormaar, J.F. 1989. Management effects on phosphorus transformation and implications for soil test recommendations. Pages 526–543 *in* Proceedings Soils and Crops Workshop 16–17 Feb. 1989. University of Saskatchewan, Saskatoon, Sask.

Teich, A.H.; Nelson, K. 1984. Survey of Fusarium head blight and possible effects of cultural practices in wheat fields in Lambton County in 1983. Can. Plant Dis. Surv. 64:11–13.

Ukrainetz, H.; Brandt, S. 1986. Wheat rotations in the Black and Gray soil zones. Pages 238–253 *in* Slinkard, A.E.; Fowler, D.B., eds. Wheat production in Canada—A review. Proceedings Canadian Wheat Production Symposium. 3–5 March 1986. Saskatoon, Sask.

Williams, J.R.; Stelfox, D. 1980. Influence of farming practices in Alberta on germination and apothecium production of sclerotia of Sclerotinia sclerotiorum. Can. J. Plant Pathol. 2:169–172.

Wright, A.T.; Coxworth, E. 1987. Benefits from pulses in the cropping systems of northern Canada. Page 108 *in* Power, J.F., ed. The role of legumes in conservation tillage systems. Proceedings Soil Conservation Society America Conference. University of Georgia, Athens, Ga. 27–29 April 1987.

Yarham, D.J. 1981. Practical aspects of epidemiology and control. Pages 353–384 *in* Asher, M.J.C.; Shipton, P.R., eds. Biology and control of take-all. Academic Press, London. 538 pp.

Zentner, R.P.; Campbell, C.A. 1988. First 18 years of a long-term crop rotation study in southwestern Saskatchewan—yields, grain protein, and economic performance. Can. J. Plant Sci. 68:1–21.

Zentner, R.P.; Campbell, C.A.; Brandt, S.A.; Bowren, K.E.; Spratt, E.D. 1986. Economics of crop rotations in western Canada. Pages 254–317 *in* Slinkard, A.E.; Fowler, D.B., eds. Wheat production in Canada—A review. Proceedings Canadian Wheat Production Symposium. 3–5 March 1986. Saskatoon, Sask.

Zentner, R.P.; Campbell, D.W.; Campbell, C.A.; Read, D.W.L. 1984*a*. Energy considerations of crop rotations in southwestern Saskatchewan. Can. Agric. Eng. 26:25–29.

Zentner, R.P.; Campbell, C.A.; Johnson, P.J.; Bacon, G. 1987*a*. Assessing the risks of re-cropping in the Brown soil zone. Agriculture Canada, Ottawa, Ont. Canadex 821.112.

Zentner, R.P.; Campbell, C.A.; Read, D.W.L.; Anderson, C.H. 1984*b*. An economic evaluation of crop rotations in southwestern Saskatchewan. Can. J. Agric. Econ. 32:37–54.

Zentner, R.P.; Campbell, C.A.; Selles, F.; McConkey, B.G.; Nicholaichuk, W.; Beaton, J.D. 1988*a*. Snow trapping and nitrogen management for zero tilled spring wheat in southwestern Saskatchewan. Pages 147–158 *in* Proceedings Great Plains Soil Fertility Workshop, vol. 2. Denver, Colo.

Zentner, R.P.; Spratt, E.D.; Reisdorf, H.; Campbell, C.A. 1987*b*. Effect of crop rotation and N and P fertilizer on yields of spring wheat grown on a Black Chernozemic clay. Can. J. Plant Sci. 67:965–982.

Zentner, R.P.; Stephenson, J.E.; Johnson, P.J.; Campbell, C.A.; Lafond, G.P. 1988*b*. The economics of wheat rotations on a heavy clay Chernozemic soil in the Black soil zone of east-central Saskatchewan. Can. J. Soil Sci. 68:389–404.

Zentner, R.P.; Stumborg, M.A.; Campbell, C.A. 1989. Effect of crop rotations and fertilization on energy balance on typical production systems on the Canadian prairies. Agric. Ecosyst. & Environ. 25:217–232.

INDEX

Note: If the page number is followed by *t*, pertinent information may be found in a table on that page. Boldface numbers refer to figure page numbers.